Prentice-Hall
Foundations of
Modern Organic Chemistry
Series

KENNETH L. RINEHART, JR., Editor

Volumes published or in preparation

MOLECULAR REACTIONS AND PHOTOCHEMISTRY

Charles H. DePuy

Professor of Chemistry
University of Colorado

Orville L. Chapman

Professor of Chemistry
Iowa State University

PRENTICE-HALL, INC., ENGLEWOOD CLIFFS, NEW JERSEY

© 1972 by Prentice-Hall, Inc.

10   9   8   7   6   5   4   3   2   1

Library of Congress Catalog Card Number 71-38244
ISBN: 0-13-599589-2 (p)
0-13-599571-X (c)
Printed in the United States of America

PRENTICE-HALL INTERNATIONAL, INC., London
PRENTICE-HALL OF AUSTRALIA, PTY, LTD., Sydney
PRENTICE-HALL OF CANADA, LTD., Toronto
PRENTICE-HALL OF INDIA (PRIVATE) LTD., New Delhi
PRENTICE-HALL OF JAPAN, INC., Tokyo

Contents

3

PHOTOCHEMICAL EXCITATION   29

4

INTRODUCTION TO PHOTOCHEMICAL
REACTIONS   44

5

STUDY OF THE MECHANISMS OF PHOTOCHEMICAL
REACTIONS   69

6

MOLECULAR ORBITAL SYMMETRY AND THE
STEREOCHEMISTRY OF CONCERTED UNIMOLECULAR
REACTIONS   84

7

CYCLOADDITION   115

Foreword

Organic chemistry today is a rapidly changing subject whose almost frenetic activity is attested by the countless research papers appearing in established and new journals and by the proliferation of monographs and reviews on all aspects of the field. This expansion of knowledge poses pedagogical problems; it is difficult for a single organic chemist to be cognizant of developments over the whole field and probably no one or pair of chemists can honestly claim expertise or even competence in all the important areas of the subject.

Yet the same rapid expansion of knowledge—in theoretical organic chemistry, in stereochemistry, in reaction mechanisms, in complex organic structures, in the application of physical methods—provides a remarkable opportunity for the teacher of organic chemistry to present the subject as it really is, an active field of research in which new answers are currently being sought and found.

To take advantage of recent developments in organic chemistry and to provide an authoritative treatment of the subject at an undergraduate level, the *Foundations of Modern Organic Chemistry Series* has been established. The series consists of a number of short, authoritative books, each written at an elementary level but in depth by an organic chemistry teacher active in research and familiar with the subject of the volume. Most of the authors have published research papers in the fields on which they are writing. The books will present the topics according to current knowledge of the field, and individual volumes will be revised as often as necessary to take account of subsequent developments.

The basic organization of the series is according to reaction type, rather than along the more classical lines of compound class. The first ten volumes in the series constitute a core of the material covered in nearly every one-year organic chemistry course. Of these ten, the first three are a general introduction to organic chemistry and provide a background for the next six, which deal with specific types of reactions and may be covered in any order. Each of the reaction types is presented from an elementary viewpoint, but in a depth not possible in conventional textbooks. The teacher can decide how much of a volume to cover. The tenth examines the problem of organic synthesis, employing and tying together the reactions previously studied.

The remaining volumes provide for the enormous flexibility of the series. These cover topics which are important to students of organic chemistry and are sometimes treated in the first organic course, sometimes in an intermediate course. Some teachers will wish to cover a number of these books in the one-year course; others will wish to assign some of them as outside reading; a complete intermediate organic course could be based on the eight "topics" texts taken together.

The series approach to undergraduate organic chemistry offers then the considerable advantage of an authoritative treatment by teachers active in research, of frequent revision of the most active areas, of a treatment in depth of the most fundamental material, and of nearly complete flexibility in choice of topics to be covered. Individually the volumes of the Foundations of Modern Organic Chemistry provide introductions in depth to basic areas of organic chemistry; together they comprise a contemporary survey of organic chemistry at an undergraduate level.

KENNETH L. RINEHART, JR.

University of Illinois

1

Introduction

The total energy of a molecule is the sum of electronic energy, vibrational energy, rotational energy, and translational energy. The first three energies are quantized;

$$E_{\text{TOTAL}} = E_{\text{ELECTRONIC}} + E_{\text{VIBRATIONAL}} + E_{\text{ROTATIONAL}} + E_{\text{TRANSLATIONAL}}$$

$$E_{\text{ELECTRONIC}} \gg E_{\text{VIBRATIONAL}} > E_{\text{ROTATIONAL}}$$

that is, they can change only by discrete jumps ($\Delta E = h\nu$). Translational energy is not quantized and changes in a continuous manner. There are two different means by which energy can be supplied to molecules. First, changing the temperature produces a continuous increase in energy. Second, a quantum of energy may be absorbed by the molecule from a beam of quanta (a light beam). These two modes of energy input give rise to thermal chemistry and photochemistry.

1.1 THERMAL ENERGY

Ordinarily we think of supplying energy to a reacting system by increasing the reaction temperature. As the temperature is raised the molecules move more rapidly; i.e. *translational energy* increases. A molecule can move with any velocity. (Translational energy is not quantized.) At a given temperature, there will be an energy distribution among the molecules, some moving much more rapidly and some much more slowly than the average. Only a few molecules will have sufficient energy to react. At a higher temperature a larger fraction of the molecules will have sufficient energy, and the rate will increase (Fig. 1-1).

As the temperature is raised, molecules can acquire additional vibrational and rotational as well as translational energy. Vibrational energy and rotational energy are quantized; that is, they can change only by discrete jumps. This can be illustrated most simply for a diatomic molecule by means of a Morse curve (Fig. 1-2). If we imagine two atoms coming together from a distance, they may eventually join to form a chemical bond, and the energy of the system will decrease. If the internuclear distance decreases below the equilibrium position, nuclear-nuclear repulsions increase rapidly, and the energy rises. The molecule thus finds itself in a "potential well" corresponding to a chemical bond.

Within this potential well the molecule can occupy any of a number of

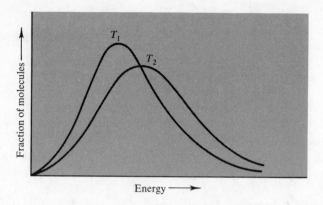

Fig. 1-1

discrete vibrational energy levels. Notice that the lowest energy level is not at the bottom of the well, and molecules will continue to vibrate even at absolute zero. When energy is supplied to the molecule, higher vibrational states ($\nu_1\nu_2$, etc.) may become populated. Only the exact amount of energy needed to go from ν_0 to ν_1, ν_2, ν_3 or some higher level may be absorbed. In typical organic molecules ν_1 lies from 2 to 10 kcal/mole above ν_0. Molecules at room temperature have an average thermal energy content of about 0.6 kcal/mole. It is thus obvious that most molecules are in their lowest vibrational energy level under these conditions. As the temperature is raised, some of the additional energy will go into populating higher vibrational levels. Many chemical reactions, especially those that are intramolecular, involve these higher vibrational levels.

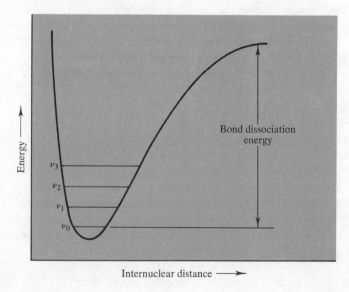

Fig. 1-2

At still higher temperatures, enough vibrational energy may be absorbed to bring about rupture of a bond. The minimum energy required to do this is known as the *bond dissociation energy* and is shown in Fig. 1-2. The amount of energy required to dissociate a bond varies widely, depending on the structure of the molecule and the nature of the atoms involved in the bond. Typical bond dissociation energies are 101 kcal/mole for the C-H bond in methane and 83 kcal/mole for the C-C bond in ethane.

1.2 ACTIVATION ENERGY

In general, energy must be supplied to molecules in order to bring about a chemical transformation. This is true even though reaction may be a favorable one in the sense that its equilibrium lies almost completely in the forward direction. The reacting molecules must first acquire enough energy to traverse the energy barrier separating reactants and products. This barrier is known as the *activation energy*† for the reaction in question. It may be so low that molecules have enough thermal energy at room

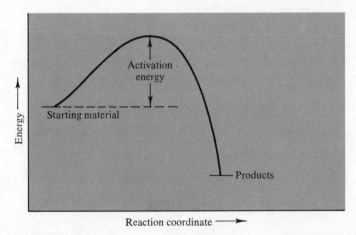

Fig. 1-3

temperature that they can cross it. Under these conditions a spontaneous reaction will occur. When the activation energy for a process exceeds that available at room temperature, additional energy must be supplied to the system before a chemical change occurs. The higher the activation energy, the slower the reaction at a given temperature. The activation energy is intimately associated with the *rate* of a chemical reaction. A reaction that occurs rapidly (i.e., has a large rate) under a given set of experimental

† For a full discussion, see W. H. Saunders, Jr., *Ionic Aliphatic Reactions* (Englewood Cliffs, N. J.: Prentice-Hall, Inc., 1965).

conditions has a lower activation energy than one that proceeds more slowly under those same conditions.† The rate of a chemical reaction of the type $B \rightarrow C$ depends on the concentration of B, the activation energy of the reaction, and the temperature. The rate of production of C will be given by the product of a rate constant k and the concentration of B.

$$\text{Rate of production of } C = \frac{dC}{dt} = k[B]$$

$$k = Ae^{-E_A/RT}$$

$$E_A = E_{\text{act}} + \tfrac{1}{2}RT$$

A is a constant

The rate constant k is a function of the activation energy and temperature. Only those molecules that have energy in excess of the activation energy will react. The fraction (N^*/N) of molecules with $E > E_{\text{act}}$ is a function of temperature. This fraction

$$\frac{N^*}{N} = e^{-E_{\text{act}}/RT}$$

is quite small. Values of N^*/N are tabulated below for a reaction with $E_{\text{act}} = 25$ kcal/mole. In the gas phase at one atmosphere, molecules collide about 10^9–10^{10} times per second. The probability of reaction is related to the product of N^*/N and the collision frequency. The small

N^*/N for	Temperature (°C)
10^{-19}	20°
10^{-15}	100°
10^{-12}	200°
10^{-10}	300°
10^{-8}	400°
10^{-7}	500°

fraction of molecules with sufficient energy is offset by the high collision frequency.

1.3 PHOTOCHEMICAL ENERGY

A second means of exciting molecules involves absorption of electromagnetic radiation.‡ The amount of energy that such radiation contains depends upon its wavelength according to the relationship given below,

---

† This statement holds for reactions with the same entropy of activation.

‡ For a description of absorption spectroscopy, see J. R. Dyer, *Applications of Absorption Spectroscopy of Organic Compounds* (Englewood Cliffs, N. J.: Prentice-Hall, Inc., 1965).

where E is the energy per molecule, h is Planck's constant, and v is the

$$E = hv$$

$$v = \frac{c}{\lambda}$$

frequency of the radiation. The frequency v and wavelength λ are inversely related. The energy of light in kcal/mole is given by the expression

$$E(\text{kcal/mole}) = \frac{2.86 \times 10^5}{\lambda(\text{Å})}$$

An energy of 1 kcal/mole corresponds to radiation of wavelength 286,000 Å or

$$\frac{10^8 \text{ Å/cm}}{286,000 \text{ Å}} = 353 \text{ cm}^{-1}$$

in the infrared portion of the spectrum.† Ten kcal/mole corresponds to 3530 cm^{-1}, also in the infrared region of the spectrum. When electromagnetic radiation of these frequencies is absorbed, molecules are excited to higher vibrational states. Radiation of shorter wavelength (higher frequency) contains more energy. Visible light has a wavelength of 4000 Å (violet) to 8000 Å (red). Light of these wavelengths contains 71–36 kcal/mole. Ultraviolet light is of even shorter wavelength (2000–4000 Å for the near ultraviolet, 100–2000 Å for the far ultraviolet). Light of 2000 Å corresponds to an energy of 143 kcal/mole. Light in the ultraviolet-visible region has energy sufficient to excite molecules to higher *electronic states*.

You are undoubtedly familiar with the idea that the absorption of light by an atom involves the excitation of an electron to a higher-energy orbital. Thus a hydrogen atom with a single $1s$ electron can absorb light and excite the electron to the $2s$ orbital. The absorption of electromagnetic radiation thus produces a transition from the ground electronic state ($1s^1$) to the first excited state ($2s^1$). When we discuss *molecules*, we need to consider *molecular orbitals* rather than atomic orbitals. Let us complete our Morse curve (Fig. 1-2). There we have drawn the energy curve that results when two atoms come together to form a bond. In Figure 1-4, the lower curve corresponds to the ground electronic state, and the upper curve to an excited electronic state. Light of the correct frequency ($\Delta E = hv$) can be absorbed by the molecule. The transition involves excitation of an electron from a bonding molecular orbital to an antibonding molecular orbital. A great deal more will be said later about the shape and properties of molecular orbitals, both bonding and antibonding, and the chemical and physical consequences of this form of

† Frequency is often expressed as waves per centimeter (cm^{-1}). There are 10^8 Å/cm.

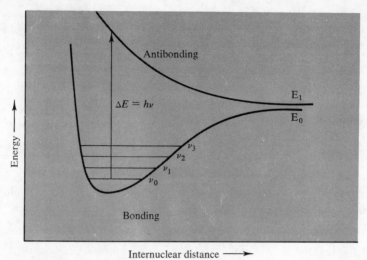

$\Delta E = h\nu$

Internuclear distance ⟶

Fig. 1-4

energy absorption. Absorption of light provides the means of introducing a large amount of energy (36–143 kcal/mole) into a molecule. Clearly, the introduction of so much energy will have profound effects on the molecule. Photochemistry is the study of the chemistry of electronically excited molecules produced by the absorption of electromagnetic radiation.

1.4 SUMMARY

Thermal excitation and photochemical excitation provide two complementary methods of introducing energy into molecules. Thermal excitation introduces energy randomly into translational, rotational, and vibrational modes, producing an energy distribution in the system such that most molecules have about the same amount of energy. Absorption of electromagnetic radiation in the visible or ultraviolet region of the spectrum excites an individual molecule instantaneously to an excited electronic state. This process involves promotion of an electron from a bonding molecular orbital to an antibonding molecular orbital. A large amount of energy is thus placed in a single molecule. It is not surprising that the chemical consequences of the two methods of introducing energy are often profoundly different.

1.5 PROBLEMS

1. How much energy does light of 1000 Å contain in kcal/mole? Of 10,000 Å? Of 1000 cm^{-1}?

2. Absorption of microwave radiation leads to changes in the rotational energy levels of a molecule. If a molecule absorbed at 10 cm^{-1}, what would be the energy separation of the rotational levels?

2

Thermal Rearrangements and Eliminations

2.1 INTRODUCTION

When an organic compound is heated, molecular motions increase. Molecules collide with one another more frequently and with greater energy. As a consequence of these collisions, vibrational motions within the molecule also increase. If we take ethane as a simple example, the vibrational motions of the carbon-carbon and the carbon-hydrogen bonds become more and more energetic. Eventually, if we raise the temperature enough, these vibrations become so energetic that there is not sufficient binding energy to hold the nuclei together; somewhere in the molecule a bond breaks, and the fragments fly apart. Ethane, for instance, may decompose on heating into two methyl radicals. The carbon-carbon bond in ethane is a strong one (83 kcal/mole), and temperatures in excess of

$$CH_3 \longrightarrow CH_3 \xrightarrow{\Delta} 2\ CH_3\cdot$$

600°C are required in order to shake it apart. In many organic molecules certain selective, interesting, and synthetically useful reactions occur at temperatures well below those required simply to break a carbon-carbon bond. In these molecules, reorganizations of the bonding electrons occur, giving rise to new molecules, which may be fragments of the larger one, or which may be new and more stable isomers of the starting material. These reactions occur at relatively low temperatures because new bonds are made simultaneously with the breaking of existing bonds in the molecule; some of the energy that is consumed in the breaking of existing bonds is regained in the formation of new ones. One particularly descriptive name for such reactions is "thermal reorganizations." The electrons and atomic nuclei reorganize themselves so that a new combination arises. In these reorganizations discrete intermediates such as free radicals, carbonium ions, and carbanions are not involved. We call these processes "molecular reactions" to emphasize that the molecule as a whole is involved and to distinguish them from ionic and free radical processes.

2.2 THE COPE REARRANGEMENT

The Cope rearrangement serves as a convenient vehicle for the discussion of molecular reactions, because it illustrates the principles of this type of transformation. The Cope rearrangement in its simplest form is shown below. When a 1,5-hexadiene is heated in solution or in the vapor phase, rearrangement occurs and a new isomeric 1,5-hexadiene is formed.

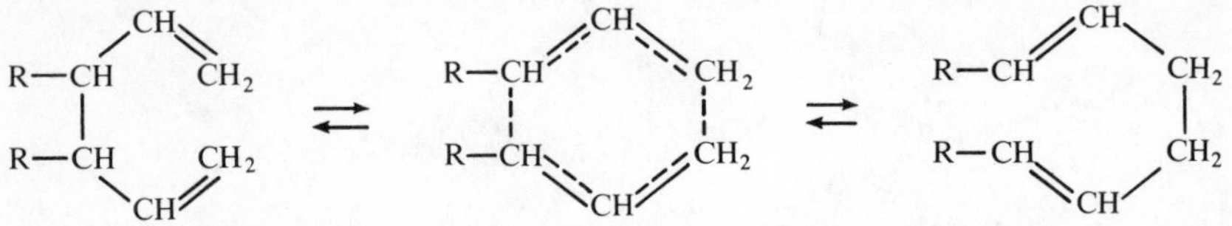

Because of this symmetry, the reaction is reversible, and an equilibrium mixture of the two hexadienes results. The temperature at which the reaction occurs is dependent upon the structure of the groups R in ways which will be discussed, but it is usually much lower than the temperature that would be required for the initial breaking of a carbon-carbon single bond. In the particular case in which the groups R are methyls, the rearrangement occurs readily at 200°.

The evidence that the Cope rearrangement is truly an intramolecular process is compelling. Suppose, for instance, that the substituted hexadienes we are considering were to break down into two allyl radicals as shown. These radicals could combine with one another to form the rearranged product. At least one other product would be expected from

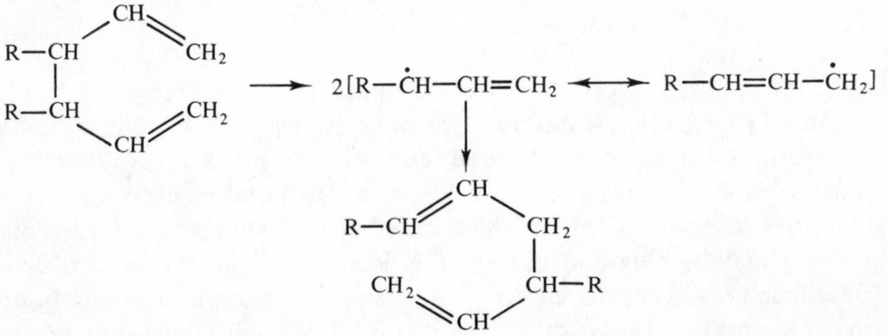

the combination of two of these radicals, the isomeric hexadiene shown. The absence of products of this type from most Cope rearrangements is a strong indication that radical intermediates are not present. Second, if a mixture of two different hexadienes is allowed to react, as, for instance, 3,4-dimethyl-1,5-hexadiene and 3,4-diethyl-1,4-hexadienes, the products after rearrangement do not contain any methyl, ethylhexadienes. If radical intermediates were involved, such "crossed products" would be expected. The absence of radical intermediates in the general case is contrasted by their observation in certain special circumstances. For instance, 1,4-diphenyl-1,5-hexadiene is one of the products from the thermal rearrangement of 3,4-diphenyl-1,5-hexadiene. The expected 1,6-diphenyl-1,5-hexadiene is also formed. In this particular case, cleavage gives a highly

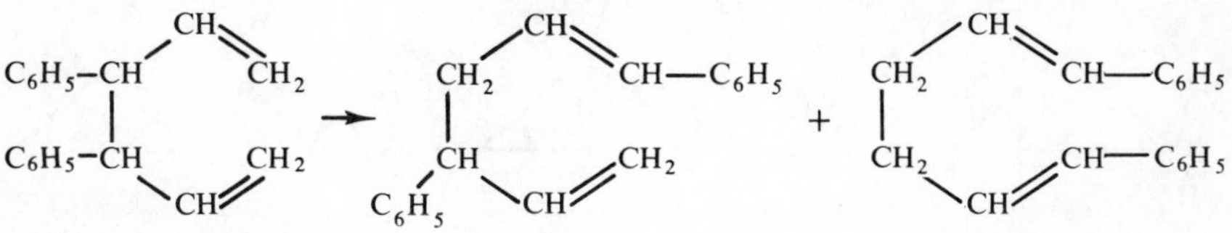

resonance stabilized radical, and, apparently, cleavage is competitive with the Cope rearrangement. In cases where this type of stabilization is absent, the highly synchronized reaction takes precedence.

Despite the fact that new bonds are made and existing bonds are broken simultaneously, there is some net loss of bonding in the transition state of the Cope rearrangement, and substituents do have an effect on the rate of the reaction by stabilizing the transition state. Phenyl groups, ester groups, and other substituents that can conjugate with the electron-deficient π system appear to have an accelerating effect. The effect is not normally as large as that which would be observed in the case of an ionic reaction, but important effects are observed.

2.3 DIVINYLCYCLOPROPANES

A striking example of acceleration, which at the same time offers additional evidence of the intramolecular nature of the reaction, is shown below. Divinylcyclopropane is a special case of a 1,5-hexadiene in which $R + R = CH_2$. Two isomers of this molecule, *cis* and *trans*, exist. The *cis* isomer has not been isolated. All attempts to synthesize it lead, even at $-40°$, to the Cope rearrangement product, 1,4-cycloheptadiene. This

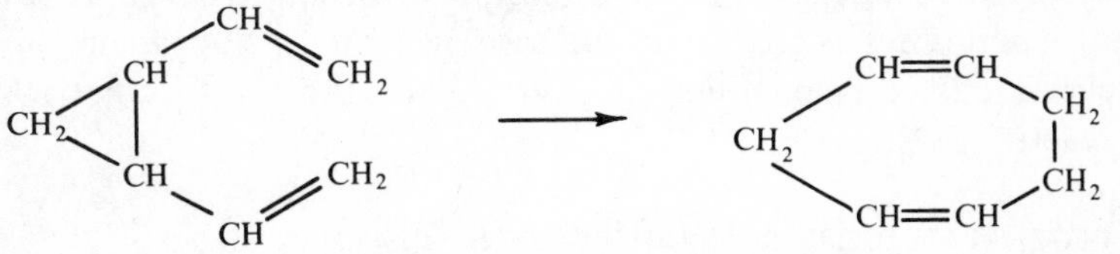

transformation is especially easy because the high strain energy of the cyclopropane ring is no longer present in the product. *trans*-1,2-Divinylcyclopropane, on the other hand, can be prepared readily. It is relatively stable to heat, although at 200° it also is transformed into 1,4-cyclohepta-diene. Notice that in the *trans* isomer it is not possible to have a concerted bond-making and bond-breaking process in the Cope rearrangement, since the ends of the 1,5-hexadiene system cannot approach one another, being constrained to opposite sides of the three-membered ring.

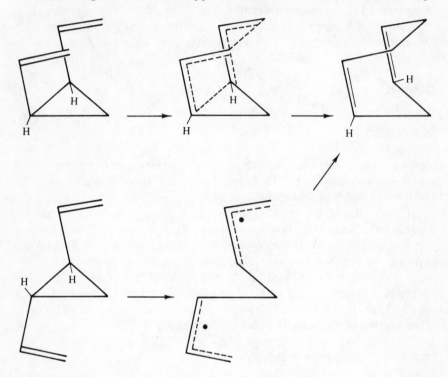

In this case the bond-breaking process must take place first, giving rise to the allylic radicals shown. These radicals, by rotation, can join at their ends, forming 1,4-cycloheptadiene. The vast difference in reactivity of the two isomers is a consequence of the lowering of the energy of activation that occurs in making the process synchronous. *cis*-1,2-Divinylcyclo-butane similarly rearranges at a relatively low temperature (120°), form-ing 1,5-cyclooctadiene. Here the amount of strain relieved in the re-arrangement is not as great as in the case of divinylcyclopropane, and the temperature is correspondingly higher. Again, the *trans* isomer is much less reactive.

2.4 DEGENERATE COPE REARRANGEMENTS

An amusing special case of the Cope rearrangement arises in a mole-cule in which the ends of the double bonds of *cis*-divinylcyclopropane are joined by a methylene group. The resultant molecule contains a strained 1,5-hexadiene system, which should be expected to undergo the Cope rearrangement readily. On the other hand, the product of the reaction is identical with the starting material, and the rearrangement is said to be "degenerate." This compound has been synthesized and its properties investigated by nuclear magnetic resonance spectroscopy.†️ Indeed, the compound does flip back and forth between these two struc-

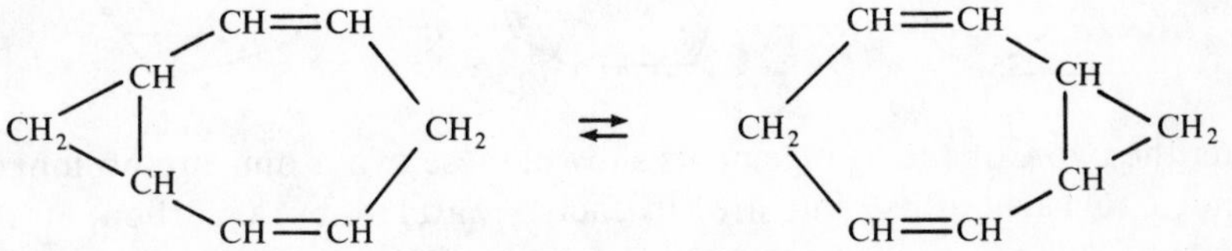

tures several times every second at room temperature. The term "fluctional molecule" has been coined to describe this unique type of behavior.

The rate at which rapid rearrangement occurs in this system can be speeded up by joining the two ends of the system together, so that the most favorable steric relationship for the reaction is maintained. The resulting molecule, called *barbaralone*, fluctuates between the two identical

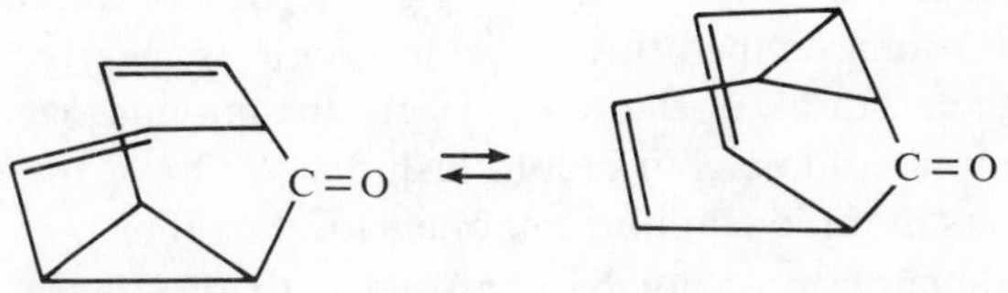

structures shown over 8000 times per second even at $-30°$.

The Cope rearrangement and the idea of a fluctional molecule reach a peak in *bullvalene*. Three different views of this highly symmetrical molecule, in which the carbonyl group of barbaralone is replaced by a

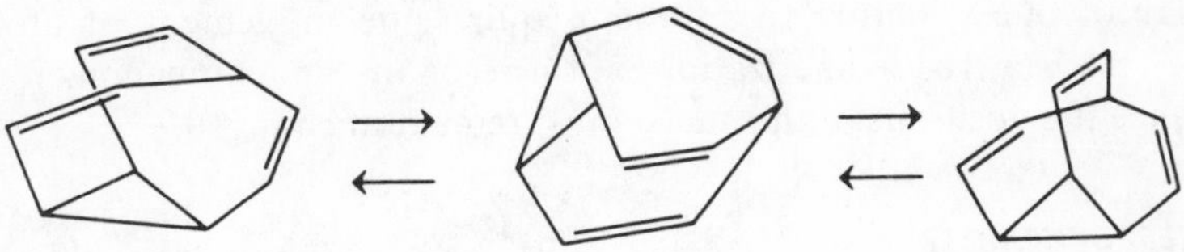

third vinyl group, are shown. Three equivalent Cope rearrangements are possible. A careful examination of the structure of Cope rearrangement products will show that in each case the product is identical to the starting material. Each of the rearrangement products can itself undergo a Cope rearrangement in three directions. One such rearrangement will, of course,

† See J. R. Dyer, *Applications of Absorption Spectroscopy of Organic Compounds* (Prentice-Hall, Inc., Englewood Cliffs, N. J.: 1965).

lead back to the starting material, but the other two lead on to new isomeric trienes, all of identical structure. The result of successive Cope rearrangements in this molecule can be quite startling. Let us examine the fate of two carbon atoms initially adjacent to one another.

After the two Cope rearrangements shown, these two atoms are no longer bonded to each other, but are, in fact, separated by a carbon atom. Further Cope rearrangements could separate them even farther. In fact, the unique properties of this system insure that no two carbon atoms remain bonded to one another very long, but that they will continually move throughout the entire molecule. As a consequence, each of the ten C-H groups can occupy any of the ten positions in the molecule. Since there are ten C-H groups and threefold symmetry, there are 10!/3 or 1,209,600 identical isomers possible for bullvalene. Bullvalene has been synthesized and shown to be rearranging at the rate of about 4000 times per second at room temperature. The molecule is nearly spherical in shape, and one can think of the ten C-H groups moving over the surface of this sphere in a completely random fashion. We have then a series of identical isomers in rapid equilibrium, one with the other.

This phenomenon must not be confused with resonance. In the case of benzene, for instance, we do not have two identical isomers in rapid equilibrium but rather a single structure with properties intermediate between those of the two Kekulé formulas.† In bullvalene this cannot be the case, because it can be shown geometrically that it is impossible to place ten groups symmetrically about the surface of a sphere. The isomerization of one structure to another requires the movement of atomic nuclei. This is a true isomerization, not a resonance phenomenon, for in resonance the positions of the nuclei must remain unchanged.

2.5 STEREOCHEMISTRY

Additional insight into the mechanism of the Cope rearrangement is gained by a consideration of the stereochemistry of the reaction. This has been studied carefully with 3,4-dimethyl-1,5-hexadienes. Notice that there are three possible geometrical isomers of the hexadiene that would

† See N. L. Allinger and J. Allinger, *Structures of Organic Molecules* (Englewood Cliffs, N. J.: Prentice-Hall, Inc., 1965).

result from this rearrangement. The actual isomers formed in the reaction were found to be dependent on the stereochemistry of the starting diene. Three stereoisomers of this diene can exist, one racemic pair and one meso isomer. The structure of this meso isomer is shown, together with the three possible ways in which it might rearrange. Since six carbon atoms are involved in the Cope rearrangement, the transition states bear

meso isomer

a close analogy to the chair and boat forms of the cyclohexane ring. Rearrangement of the *meso*-diene through a boat-type transition stage would give either the *cis,cis* or the *trans,trans* product. If a chair-type transition state were favored, then the *cis,trans* isomer would result. When the reaction was actually carried out, 99.7% *cis,trans*-2,6-octadiene was actually formed along with 0.3% of the *trans,trans* isomer. No *cis,cis* isomer could be detected. Obviously, a chair-type transition state is highly favored over a boat-type, just as a chair form of cyclohexane ring is favored over the boat form.

In the case of the rearrangement of the racemic mixture, the situation is slightly more complicated. Here, two different chair-type transitions states are possible, one leading to the *cis,cis*, and the other to the *trans, trans* isomer. When the racemic mixture rearranged, 90% of the *trans,trans* and 9% of the *cis,cis* isomer were formed. Less than 1% of the *cis,trans*

CH$_3$
H
H
CH$_3$

CH$_3$
H
CH$_3$
H
CH$_2$
CH$_2$

cis, cis

H
CH$_3$
CH$_3$
H

CH$_3$
CH$_2$
H H
CH$_3$
CH$_2$

S,S-isomer of racemic
mixture

trans, trans

isomer, the product of a boat-type transition state, could be detected among the products, showing that the chair-type transition state is favored. The predominance of the *trans,trans* isomer over the *cis,cis* is also to be expected, since the former is the more stable of the two products. These stereochemical results are important in synthesis, and they also provide additional evidence that the reaction does not involve free radical intermediates, since great stereoselectivity would not be expected from the recombination of substituted allyl radicals.

Despite the favorability of the chair over the boat transition state (about six kilocalories per mole), it is clear that the Cope rearrangement can occur through a boat transition state if no other option is open to it. This is the case, for instance, in the rearrangement of *cis*-1,2-divinyl-cyclopropane, since a chair transition state would lead to the highly strained *cis, trans*-1,4-cycloheptadiene.

2.6 THE CLAISEN REARRANGEMENT

The Cope rearrangement is an example of a general type of rearrangement. Carbon, nitrogen, and oxygen atoms may be used almost interchangeably in any of the positions. In the Claisen rearrangement, atom *A* or *B* is oxygen; the others are carbon. The rearrangement of such a system leads to an unsaturated carbonyl compound. Historically, the first example of this rearrangement was discovered when the allyl ether of the enol form of acetoacetic ester was heated. Later, the more important observation was made that the allyl ether of phenol rearranged in a similar manner when heated. In this case one of the double bonds of the hexadiene system is part of a benzene ring, which has the result of making the temperature required for the reaction somewhat higher. A great

$$D \overset{A-B}{\underset{E \quad G}{\diagup \quad \diagdown F}} \quad \rightleftharpoons \quad D \overset{A \quad B}{\underset{E-G}{\diagup \quad \diagdown F}}$$

$$\underset{\underset{CO_2C_2H_5}{|}}{\overset{CH_3}{\underset{HC}{C}}}\!\!O^- \;+\; \underset{CH_2}{\overset{CH_2Br}{CH}} \;\rightarrow\; \underset{\underset{CO_2C_2H_5}{|}}{\overset{CH_3}{C}}\!\!\overset{O-CH_2}{\underset{CH \quad CH_2}{}}CH \;\rightarrow\; \underset{\underset{CO_2C_2H_5}{|}}{\overset{CH_3}{C}}\!\!\overset{O}{\underset{CH-CH_2}{}}\!\!\overset{CH_2}{CH}$$

amount of work has been carried out on this reaction, and all of it is in accord with the mechanism written below. If, for instance, a substituted allyl ether is used in the rearrangement, the position of the substituents is found to be inverted from that of the starting material, and careful experiments have been carried out using radioactive tracers in order to

prove that the fragments do not separate in the course of the rearrangement. In most details, then, the Claisen rearrangement is quite similar

to the Cope rearrangement.

2.7 THE *PARA* CLAISEN REARRANGEMENT

A particularly interesting Claisen rearrangement is observed in the case of allyl phenols with methyl groups in both *ortho* positions. The result is the *para Claisen rearrangement*, in which the allyl group appears, after rearrangement, not in the *ortho* but in the *para* position. Again the process passes all the tests for an intramolecular process, but now if a substituted allyl group is used, the substituent occurs in the product in the same relative position that it did in the starting material. Two possibilities are evident. Either the carbon has passed from the oxygen to the *para* position without any inversion of the structure of the allyl group, or two inversions have occurred, which would result in a net overall retention of the side chain structure.

The latter of these two possibilities actually occurs in the *para* Claisen rearrangement. First, a normal *ortho* Claisen rearrangement takes place, leading to the *ortho*-substituted cyclohexadienone. Because of the presence of both the methyl and allyl group on the *ortho* carbon, this intermediate cannot aromatize. The molecule does, however, contain a 1,5-hexadiene system, so the initial Claisen rearrangement is followed by a Cope rearrangement, resulting in substitution at the *para* position. Since this position does contain a hydrogen atom, the molecule readily converts itself to the stable phenyl derivative.

A number of lines of evidence support this pathway for the *para* Claisen rearrangement. The intermediate cyclohexadienone has been synthesized independently and shown to be converted readily into the *p*-allyl phenol on heating. An ingenious series of experiments was carried out with allyl 2,6-diallylphenyl ether tagged with radioactive carbon as shown. Rearrangement to the *ortho* position produces a cyclohexadienone in which either the labeled or unlabeled allyl group can be transferred to the *para* position. The amount of radioactivity at the *para* position and its position in the allyl group were fully in accord with the pathway proposed for the *para* Claisen rearrangement.

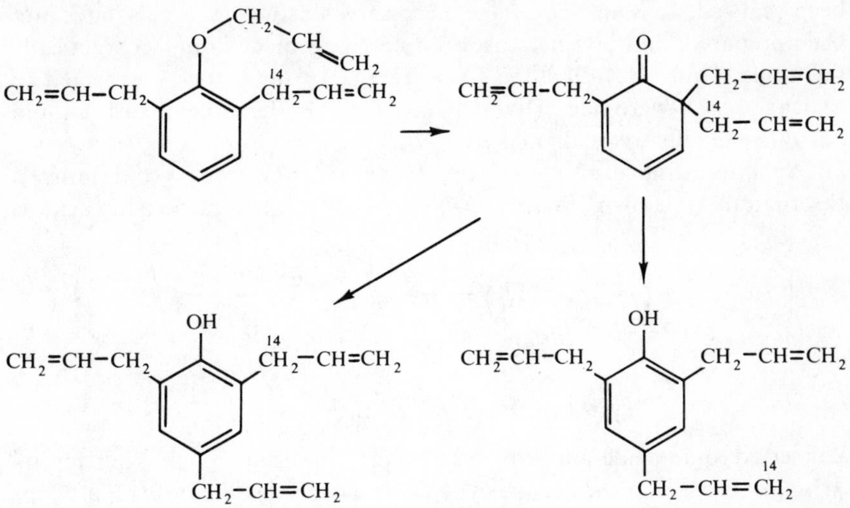

A systematic study has been made of the rates of rearrangement of *para*-substituted phenyl allyl ethers in an attempt to probe into the electronic requirements of the Claisen rearrangement. In general, *ortho,para* directing groups like methoxyl groups and halogen speed up the reaction, whereas *meta* directing groups have little or no effect on the rate. None of the substituents has a large effect on the rate of the reaction in accord with the highly concerted thermal reorganization mechanism.

The Claisen rearrangement has important synthetic applications. It is difficult to synthesize molecules containing quaternary carbon centers. Many naturally occurring substances contain such centers, and methods by which such centers can be generated attract considerable interest. An example of the utility of the Claisen rearrangement for this purpose is shown below. A vinyl ether of a β,γ-unsaturated alcohol on heating rearranges to an aldehyde.

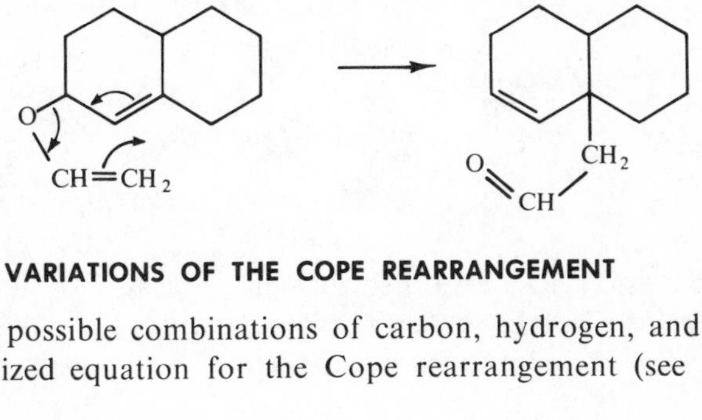

2.8 OTHER VARIATIONS OF THE COPE REARRANGEMENT

Not all possible combinations of carbon, hydrogen, and nitrogen in the generalized equation for the Cope rearrangement (see p. 15) have been realized. In some cases the necessary starting materials have not been prepared, and in other cases the rearrangement is not energetically favorable. Thus a 1,4-diketone (A = F = O) would not be expected to rearrange to a peroxide. Divinyl peroxide, on the other hand, should rearrange readily to the dione.

An interesting example of the Cope rearrangement containing a heteroatom is shown. When *cis*-2-vinylcyclopropanecarboxylic acid is

converted to an azide and subjected to the conditions of the Curtius rearrangement, the expected isocyanate is not isolated. Instead a Cope rearrangement takes place, leading to the amide shown. Additional examples of the Cope rearrangement involving heteroatoms are illustrated in the problems at the end of the chapter.

2.9 CYCLIC ELIMINATION REACTIONS

A number of organic molecules, including esters, amine oxides, and even some hydrocarbons, undergo thermal unimolecular decompositions with the formation of alkenes. Synthetically these reactions are of extreme importance, since they provide alternatives to the acidic or basic conditions usually employed in elimination reactions. They also provide additional examples of molecular reactions, since they appear not to involve intermediate ions or radicals.

2.10 ESTER PYROLYSIS

Most esters that contain β-hydrogen atoms decompose at 400–550° with the formation of an alkene and a carboxylic acid. The reaction may be illustrated in the simplest case with ethyl acetate.

Ethylene and acetic acid are formed in high yield at 550°. Two characteristics of the reaction account for its synthetic utility. It is highly stereospecific (*cis*), and the alkene formed is seldom extensively rearranged.

Olefins which would not survive acidic dehydrating conditions can be prepared by thermal eliminations.

Pyrolysis is also a useful alternative to basic hydrolysis of esters when hydrolysis is sterically hindered.

In a typical secondary or tertiary aliphatic ester, more than one olefinic product may be formed. Thus, *sec*-butyl acetate can form 1-butene, and *cis*- and *trans*-2-butenes on pyrolysis. A mixture of products is always formed, and the product distribution approaches that expected statistically. In the case of *sec*-butyl acetate, there are three primary hydrogens,

$$CH_3CHCH_2CH_3 \longrightarrow CH_2{=}CHCH_2CH_3 + CH_3CH{=}CHCH_3$$
$$\underset{OAc}{|} \qquad\qquad 57\% \qquad\qquad\qquad 43\%$$

$$\underset{\underset{OAc}{|}}{\overset{\overset{CH_3}{|}}{CH_3CCH_2CH_3}} \longrightarrow \underset{\underset{CH_3}{|}}{CH_2{=}CCH_2CH_3} + \underset{\underset{CH_3}{|}}{CH_3C{=}CHCH_3}$$
$$\qquad\qquad\qquad 75\% \qquad\qquad\qquad 25\%$$

any one of which may be lost with the formation of 1-butene. There are two hydrogens whose loss would lead to 2-butene. On a strictly statistical basis we would calculate a 3:2 or 60:40 ratio of these two alkenes. The ratio actually found is 57:43. Similarly, *tert*-amyl acetate would be calculated on a statistical basis to form 2-methyl-1-butene and 2-methyl-2-butene in the ratio of 6:2 or 75:25, and that exact ratio is found. This close correspondence between statistical and experimental olefin ratios is something of an oversimplification, since the ratio of *trans* to *cis*-2-butene formed is nearly 2:1, far from the statistical ratio and approaching the equilibrium ratio at the temperature involved. In cyclic systems, too, olefin mixtures will ordinarily result unless elimination in one direction is prohibited by the absence of a *cis* hydrogen atom.

Substituents have only a small effect on the direction of pyrolysis of an ester. Activation of the hydrogen by an adjacent phenyl group, for

example, results in only a modest increase in elimination in that direction. Strong acidification of the hydrogen by an adjacent carbonyl group, on the other hand, has a more pronounced effect, and leads to the exclusive formation of the conjugated ester. Substituents attached to the same carbon as the ester group being eliminated may have a small effect on the

ease with which an elimination proceeds. For primary acetates, a temperature over 500° is usually required for reaction, whereas for secondary acetates 450° is required, and for tertiary acetates, 400° suffices. The effects of substituents on the rate of the reaction are quite small in comparison with their effects on ionic reactions, and fully in accord with the hypothesis of a "molecular reaction."

2.11 THE ENE REACTION

The acetate pyrolytic elimination reaction may be considered as a special case of a general transformation known as the *ene* reaction. The simplest possible example could be imagined to occur between ethylene and propylene. There appears to be little evidence that the reaction does occur in either direction in this simplest case, but numerous examples are known in more complex systems. An example of a retro-ene reaction, that is to say the cleavage of an alkene to a diene, occurs upon heating *trans* cyclooctene or *trans*-cyclononene. In these cases the strain in the system is relieved in opening the ring, facilitating the cleavage.

The forward reaction apparently occurs readily with acetylenes.

An ester pyrolysis is, then, merely a retro-ene reaction with two carbon atoms of the system replaced by oxygen atoms. Since these thermal reactions appear to proceed through highly concerted transition states in

which there is relatively little charge separation, it is not surprising that the carbon atoms can be replaced by other atoms almost at will. In the following section we shall discuss a few examples.

2.12 PYROLYSIS OF HOMOALLYL ALCOHOLS

Homoallyl alcohols smoothly decompose upon heating to form a carbonyl compound and an alkene. This system is particularly instructive, because the reaction is frequently reversible in practice as well as in theory. The reaction of α-pinene with ethyl pyruvate is a case in point, where an equilibrium mixture of products has been approached from each side. The results in this case are also useful in confirming the

concerted nature of the process. If it were a stepwise ionic reaction, rearrangement of the pinene skeleton would be observed. Such rearrangement of the four-membered ring is a common feature of acid-catalyzed carbonium ion processes in this system.

The reaction of formaldehyde with an olefin is known as the Prins reaction, and has gained industrial importance. The rate of the reaction is accelerated by protonic or Lewis acids, and the mechanism appears to vary depending upon the nature of the alkene component and the experimental conditions.

2.13 KETONE ENOL-ETHER EQUILIBRIA

Enol ethers are readily converted to ketones under pyrolytic conditions; the temperatures required are lower than for the corresponding esters. Since a carbonyl compound is more stable than its corresponding

enol, the reverse reaction has not been observed. For the same reason, a ketone cannot be pyrolytically decomposed into an alkene and an enol.† As with previous examples, the reaction does occur if it can be provided with a suitable driving force such as the opening of a cyclopropane ring.

In keeping with the concerted, cyclic nature of the reaction, only *cis*-1-acetyl-2-methylcyclopropane reacts. The *trans* isomer is stable to the reaction conditions.

† This reaction does occur photochemically.

A pyrolytic version of a reverse aldol reaction is also known.

$$CH_3-\underset{\underset{\displaystyle OH}{|}}{CH}-\underset{\underset{\displaystyle CH}{\diagdown}}{CH_2}\overset{O}{\diagup} \quad \xrightarrow{\Delta} \quad CH_3-\underset{\underset{\displaystyle O}{\|}}{CH} \quad \underset{\underset{\displaystyle CH_2}{\diagup}}{CH}\diagdown^{HO}$$

2.14 THE ENE REACTION WITH SULFUR AND NITROGEN

Ene reactions are not limited to systems involving carbon and oxygen; historically the Chugaev reaction preceded the acetate pyrolysis as a method for the formation of alkenes from alcohols. Amides will also

undergo pyrolytic eliminations, but the temperatures required to bring about the reaction are too high for the method to have synthetic utility. In this reaction an amide is converted into an unstable tautomer. Reaction

temperatures decrease markedly if the oxygen and nitrogen atoms are interchanged, since a high-energy imino ether is transformed into an amide.

$$
\underset{\substack{\displaystyle |\\ CH_2 \text{——} CH_3}}{\overset{\substack{CH_3\\ \displaystyle |\\ C\\}}{\underset{O \diagdown \diagup NR}{}}} \xrightarrow{350°} \underset{CH_2 \text{===} CH_2}{\overset{\substack{CH_3\\ \displaystyle |\\ C\\}}{\underset{O \diagdown \diagup \overset{NR}{\underset{H}{}}}{}}}
$$

2.15 OTHER PYROLYTIC ELIMINATIONS

The Cope elimination is an extremely mild pyrolytic method for converting an amine to an alkene. The process has been shown to involve a

$$
\underset{CH_3CH_2CHCH_3}{\overset{\substack{N(CH_3)_2\\ \displaystyle |}}{}} \xrightarrow{H_2O_2} \underset{CH_3CH_2 - CH - CH_3}{\overset{\substack{\overset{-}{O} - \overset{+}{N}(CH_3)_2\\ \displaystyle |}}{}} \xrightarrow{\Delta}
$$

$$
\begin{aligned}
& CH_3CH = CHCH_3 \quad (33\%)\\
+ \; & CH_3CH_2CH = CH_2 \quad (67\%)\\
+ \; & HON(CH_3)_2
\end{aligned}
$$

cyclic transition state resembling that for the ene reaction, but with one atom less in the cycle. Reaction temperatures vary from about 100° in aqueous solutions down to room temperature and below in dimethyl

$$
\underset{\substack{\displaystyle |\\ -\overset{|}{C} \diagup \quad \overset{|}{C} \diagdown \\ \diagup \qquad \diagdown}}{\overset{\substack{\overset{\displaystyle |}{\underset{\oplus}{-N}} \text{——} \overset{\ominus}{O}\\[4pt] H}}{}} \longrightarrow \quad \underset{-\overset{|}{C} ==== \overset{|}{C}-}{\overset{\substack{-N \text{——} O\\[4pt] H}}{}} \longrightarrow \quad \underset{\diagup C = C \diagdown}{\overset{\substack{-N \text{——} O\\ \diagdown \\ H}}{}}
$$

sulfoxide solution. The corresponding sulfoxides also undergo this reaction readily, but phospine oxides do not.

2.16 PYROLYSIS OF HALIDES

Organic halides undergo a cyclo-elimination when heated in the gas phase at about 350°. The reaction lacks the laboratory utility of the ester or Cope elimination, since it is difficult to prevent competing free radical

$$
\underset{\substack{\displaystyle |\\ Br}}{\overset{\substack{CH_3\\ \displaystyle |}}{CH_3 - C - CH_3}} \xrightarrow{\Delta} \underset{\substack{\displaystyle |\\ Br \text{--} H}}{\overset{\substack{CH_3\\ \displaystyle |}}{CH_3 - C === CH_2}} \longrightarrow \underset{BrH}{\overset{\substack{CH_3\\ \displaystyle |}}{CH_3 - C = CH_2}}
$$

reactions, but it has been extensively investigated and remains of theoretical interest. Again the stereochemistry of the reaction is *cis*, and the reaction is unimolecular, as demanded by the proposed mechanism. Substituents, particularly those attached to the C-Br group, have a much greater effect upon the rate of this cyclo-elimination than they do on either the amine oxide or ester pyrolysis. Tertiary halides, for example, react about 10^4 times faster than primary halides, a factor which, extrapolated to room temperature, compares favorably with the relative rates of solvolysis of these compounds in acetic acid. It is clear that there is a great deal more polar character to the transition state of a halide cyclo-elimination than in either an ester or amine oxide elimination. This may be due, in large part, to the small size of the cycle in the transition state for the halide elimination, three atoms plus hydrogen. The product distribution obtained in the pyrolysis of an unsymmetrical halide is also quite different from that obtained from the corresponding ester or amine oxide.

$$CH_3CHCH_2CH_3 \longrightarrow CH_2{=}CHCH_2CH_3 + CH_3CH{=}CHCH_3$$

$$\overset{|}{Br} \qquad\qquad\qquad 40{\text -}16\% \qquad\qquad\qquad 60{\text -}80\%$$

2.17 PROBLEMS

1. Show that the following transformations are examples of the Cope rearrangement.

2. What product would you expect if the following systems underwent ene or retro-ene reactions?

3. The following reaction has been observed.

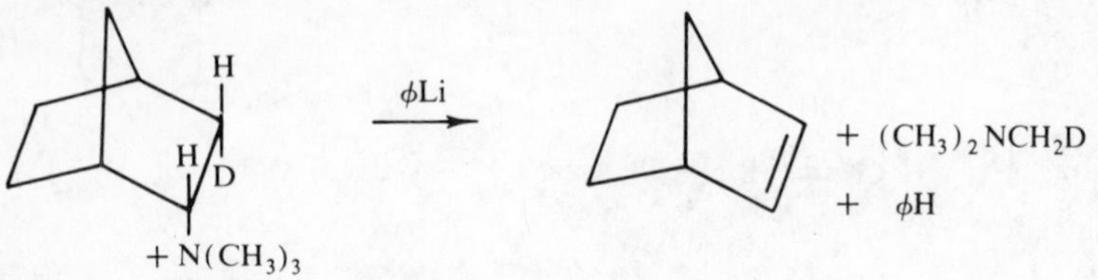

$$+ \ N(CH_3)_3$$

Give a mechanism using as analogy one of the reactions of this chapter.

4. Show that the following transformation bears the same relationship to the Cope rearrangement that the Cope elimination does to the ene reaction.

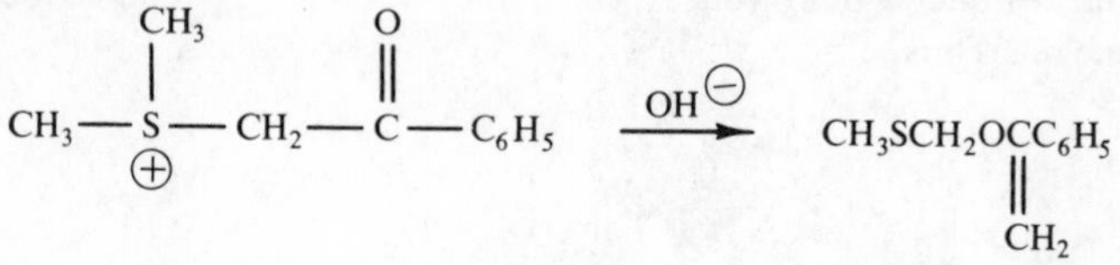

5. Show that the following reactions may be considered as a retro-ene reaction with the cyclopropane ring taking the place of a double bond.

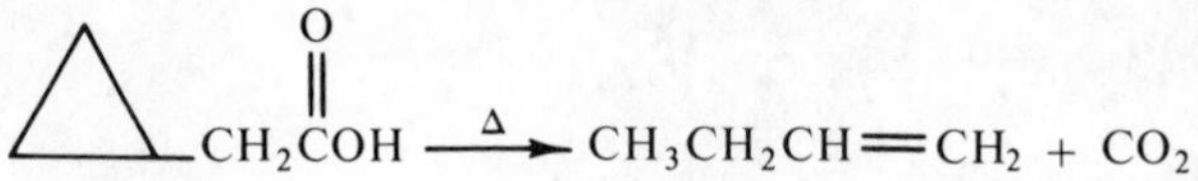

REFERENCES

Cope Rearrangement

1. A. Jefferson and F. Scheinmann, "Molecular Rearrangements Related to the Claisen Rearrangement," *Quart. Rev.*, **22** (1968), 391.

2. S. J. Rhoads, in *Molecular Rearrangements*, ed. P. deMayo, Pt. I. New York: Interscience, 1963.

Pyrolytic Eliminations

1. Allan Maccoll, "Gas-Phase Heterolysis," in *Advances in Physical Organic Chemistry*, ed. V. Gold, Vol. 3. New York: Academic Press, 1965, p. 91.

2. C. H. DePuy and R. W. King, "Pyrolytic *cis*-Eliminations," *Chem. Rev.*, **60** (1960), 431.

Ene Reaction

1. H. M. R. Hoffmann, *Angew. Chem. Internat. Ed. Engl.*, **8** (1969), 556.

Photochemical
Excitation

3.1 INTRODUCTION

The concept of excitation of molecules by absorption of ultraviolet light was introduced in Chapter 1. We now turn our attention to the practical problems of photochemical excitation and to the nature of the changes in electronic structure of molecules produced by light absorption.

3.2 LIGHT ABSORPTION

Photochemistry begins with absorption of light in the 180–800 nm (1800–8000 Å) region of the spectrum, generally termed ultraviolet light. In order to know what wavelength light we should employ in a particular photochemical experiment, we must determine the ultraviolet absorption spectrum† of the molecule we wish to study. Such a spectrum measures the amount of incident light absorbed by the molecule as a function of wavelength. The fraction of light absorbed (I/I_0) is given by the Lambert-Beer law. The molar extinction coefficient ϵ is a property of the individual compound and measures the absorptivity. Ultraviolet spectra are usually plotted as ϵ or log ϵ *vs* wavelength. The log ϵ plots are particularly useful for exhibiting weak and strong absorption bands on the same scale: Fig. 3-1 shows the ultraviolet spectrum of a typical aromatic ketone, benzophenone $(C_6H_5)_2C{=}O$. The ultraviolet spectrum of 1,3-butadiene is shown in Fig. 3-2. If we wish to excite these molecules, we must irradiate them with light in regions where they absorb. We must, therefore, match the emission of our source, usually a mercury arc lamp, to the absorption of the compound. The principal emission lines of mercury arc lamps in this region are shown superimposed on the ultraviolet spectra of benzophenone and 1,3-butadiene. Low-pressure mercury arc lamps are good sources of the 2537 Å line of mercury, and high-pressure mercury arc lamps are good sources of the 3130 and 3660 Å lines. Filters are available which permit selection of either of these lines.

---

† See J. R. Dyer, *Applications of Absorption Spectroscopy of Organic Compounds* (Englewood Cliffs, N. J.: Prentice-Hall, Inc., 1965).

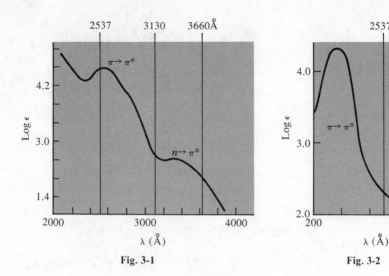

Fig. 3-1   Fig. 3-2

$$\log(I_0/I) = \epsilon cl$$

I_0 = intensity of light before entering sample
I = intensity of light after passing through sample
c = concentration in moles per liter
l = path length of light through sample in centimeters
ϵ = molar extinction coefficient

3.3 EXPERIMENTAL TECHNIQUES

The experimental techniques used in photochemistry vary widely, depending on the nature of the work. The two most common techniques used for preparative purposes in organic photochemistry are internal and external irradiation. Internal irradiation uses a mercury arc lamp protected by a water-cooled immersion well. The solution to be irradiated surrounds the immersion well and is exposed to the full output of the lamp. It is important that the material for the construction of the immersion well be selected with due consideration for its light transmission characteristics. Pyrex glass transmits most of the incident light above 3000 Å, and quartz transmits light above 2200 Å. Photochemical experiments with the 3130 and 3660 Å lines of mercury can be carried out with Pyrex glass immersion wells, whereas experiments using 2537 Å light require quartz vessels.

External irradiation may take several forms. One convenient device is the Srinivasan-Griffin reactor (Fig. 3-4). This reactor uses any one of three sets of lamps that have maximum emission at 2537 Å, 3000 Å, and 3500 Å. The latter two are low-pressure mercury arc lamps with phosphor coatings that emit at 3000 and 3500 Å, respectively. The phosphors give

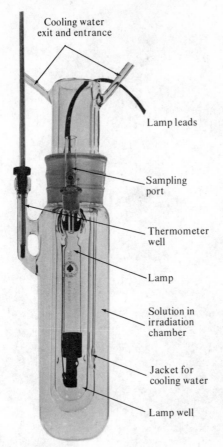

Cooling water
exit and entrance

Lamp leads

Sampling
port

Thermometer
well

Lamp

Solution in
irradiation
chamber

Jacket for
cooling water

Lamp well

Fig. 3-3 Photochemical reaction vessel. Courtesy of Ace Glass Incorporated and Englehard Hanovia.

broad band rather than sharp line emission. The vessel to be irradiated is surrounded by the lamps and is cooled by a fan or by an internal cooling coil.

3.4 ELECTRONIC TRANSITIONS

We shall examine in more detail the electronic transitions induced by light absorption. Figure 3-5 shows schematic Morse potential curves for the ground electronic state and one excited electronic state of a polyatomic molecule. This excited state potential curve has a minimum in it, so the molecule will not fly apart on excitation. The minimum in the potential curve of the excited state occurs at a larger internuclear distance than that of the ground state. This is reasonable, since the excited state has one electron in an antibonding orbital and the bonding will be weaker

Fig. 3-4 Rayonet Photochemical Reactor. Courtesy The Southern New England Ultraviolet Company.

in the excited state. There is a series of vibrational energy levels super-imposed on the potential curve for each electronic state.

Consider now the Morse curves for the ground state (E_0) and first excited state (E_1) shown in Fig. 3-5. At room temperature we have in-sufficient energy to populate excited vibrational levels, and most transitions will start from ν_0 of the ground state. The Franck-Condon principle tells us that absorption of a quantum of light will occur rapidly, even with respect to molecular vibrations. No changes in nuclear positions occur during the excitation process; i.e., we have vertical excitation as shown in Fig. 3-5. Changes in molecular structure (bond angles and bond lengths) will occur as the electronically excited molecule comes to thermal

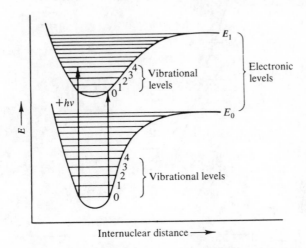

Fig. 3-5

equilibrium with its surroundings. Transition from ν_0 of the ground state may terminate in any of several vibrational levels of the excited state. This is the reason for band spectra rather than sharp lines in ultraviolet spectra. The energy of the electronic transition is measured from ν_0 of the ground state to ν_0 of the excited state, $E_1(\nu_0) - E_0(\nu_0) = \Delta E = h\nu$.

3.5 JABLONSKI DIAGRAMS

Polyatomic molecules have polydimensional energy surfaces too complex to represent adequately by simple Morse curves. Jablonski diagrams provide a useful means of representation of the excited states of such molecules. In order to appreciate a Jablonski diagram it is necessary to understand the concept of spin multiplicity. Ordinary organic molecules have an even number of electrons. In the ground state, most organic molecules have all electrons paired. Molecular states with all electrons paired are called *singlet states* (S_n). Absorption of light occurs without spin inversion,† and the initial excited state produced is a singlet excited state. Singlet excited states may undergo spin inversion, giving a new ex-

---

† Transitions in which spin inversion occurs are known, but can be observed only by use of special techniques.

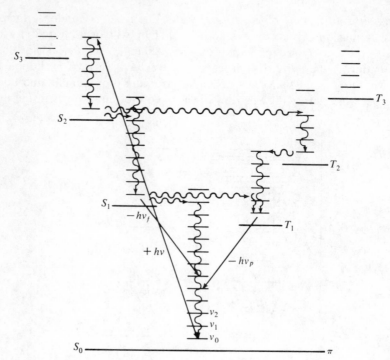

Fig. 3-6

cited state with two unpaired electrons. Molecular states with two un-paired electrons are called *triplet states* (T_n). The singlet and triplet designations derive from the fact that singlet states do not split in a mag-netic field, whereas triplet states split into three different energy levels. Free radicals that have one unpaired electron give rise to two energy levels in a magnetic field and are thus known as *doublet states*. Electronic transitions between states of the same multiplicity, i.e., singlet-singlet and triplet-triplet transitions, are spin-allowed, and transitions between states of different multiplicity, i.e., singlet-triplet or triplet-singlet transitions, are spin-forbidden. Absorptions due to spin-forbidden transitions can be observed only by the use of special techniques. Spin-forbidden transitions between excited singlet and triplet states (intersystem crossing) occur in many molecules.

The major events that occur following light absorption may be sum-marized in terms of a Jablonski diagram. Imagine absorption of a quan-tum of light of sufficient energy to induce a transition from S_0 to S_2, as shown in Fig. 3-6. In solution, the excess vibrational energy of S_2 will be rapidly dissipated by radiationless processes (vibrational cascade) to the solution. The S_2 state will undergo radiationless internal conversion to an upper vibrational level of S_1, which in turn will rapidly lose its excess vibrational energy to the medium. All of these processes will occur in about 10^{-11} seconds; i.e., the lifetime of upper excited singlet states is generally less than 10^{-11} sec. The thermally equilibrated low-lying singlet excited state $S_1(\nu_0)$ has a relatively long lifetime ($\sim 10^{-8}$ sec). The lifetime of this state is limited by four important processes: (1) fluorescence, (2) chemical reaction, (3) radiationless decay to the ground state, and (4) intersystem crossing. Fluorescence is the emission of light from a singlet excited state as it returns to the ground state. Fluorescence is, in a sense, the opposite of the lowest-energy singlet-singlet light absorption process. Nondissociative chemical reactions are more probable in long-lived excited states. For this reason, S_1 and T_1 are the major reactive states in photochemical processes. Radiationless transition from S_1 to S_0 may be thought of as internal conversion from S_1 to vibrationally excited S_0 followed by vibrational cascade to $S_0(\nu_0)$ with the medium absorbing the excess thermal energy. Intersystem crossing involves spin inversion and gives rise to a lower-energy triplet state. This process is very important in photochemistry, because the triplet state T_1 produced is even longer-lived than S_1. Low-lying triplet states in general have lifetimes greater than 10^{-6} sec, and molecules with triplet lifetimes longer than one second are known. On the other hand, upper triplet excited states have very short lifetimes, just as upper singlet states do. The lifetime of the T_1 state is limited by (1) phosphorescence, (2) chemical reaction, and (3) radiationless decay to S_0. Phosphorescence is the emission of light from a triplet state as it returns to the ground state. Phosphorescence and internal conversion of T_1 to S_0 are spin-forbidden processes, and this contributes to the relatively long lifetime of low-lying triplet excited states. Chemical reaction, especially intermolecular reaction, is favored by the longer lifetime of the T_1 state relative to the S_1 state.

3.6 INTERSYSTEM CROSSING

Intersystem crossing ($S_1 \rightarrow T_1$) is formally a spin-forbidden process. In some molecules, however, it occurs with 100% efficiency, whereas in others it does not occur to any measurable extent. The efficiency in intersystem crossing depends, among other factors, on the difference in energy between the low-lying singlet and triplet excited state (the S_1–T_1 energy gap). When this energy difference is small, there is considerable overlap of the two excited states and spin identity is less apparent. In such systems intersystem crossing is efficient. When the energy difference is large, spin-forbiddenness is quite important and intersystem crossing efficiency is low or zero. Generally speaking, ketones have high intersystem crossing efficiencies, aromatic compounds have intermediate to high intersystem crossing efficiencies, and olefins have low intersystem crossing efficiencies. The presence of heavy atoms (sulfur, chlorine, bromine, iodine, etc.) in a molecule enhances intersystem crossing efficiency. Jablonski diagrams are given in Fig. 3-7 for benzophenone, which intersystem crosses with 100% efficiency, and 1,3-butadiene, for which the intersystem crossing efficiency approaches zero.

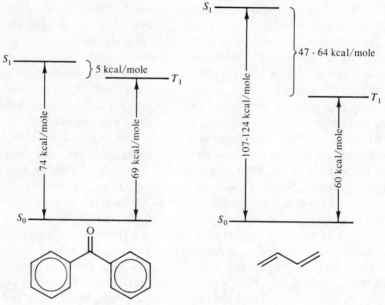

Fig. 3-7

3.7 ENERGY TRANSFER

We now need to consider a secondary means for producing electronically excited molecules. The method involves transfer of excitation energy from an electronically excited molecule to the ground state of another molecule, and is generally used for producing triplet excited states. Singlet excitation energy can be transferred, but the probability is limited by the lifetimes of excited singlet states ($\sim 10^{-8}$ sec), which are relatively short compared to the lifetimes of triplet excited states ($>10^{-6}$ sec). Triplet energy transfer in its simplest form requires that the triplet energy [$E_T = E(T_1) - E(S_0)$] of the donor be 3 kcal/mole or more greater than the triplet energy of the acceptor molecule. If this condition is met, triplet energy transfer will occur at every collision between a triplet excited donor molecule and a ground state acceptor molecule. Such reactions are said to be diffusion-controlled reactions and have pseudo first-order rate constants of 10^9 to 10^{10} sec^{-1}. Imagine a donor-acceptor system such that only the donor absorbs the incident light, and the triplet energy of the donor is at least 3 kcal/mole greater than the triplet energy of the ac-

$$D + h\nu \longrightarrow {}^1D$$

$${}^1D \xrightarrow[\text{crossing}]{\text{intersystem}} {}^3D$$

$${}^3D + A \longrightarrow D + {}^3A$$

$${}^3A \longrightarrow \text{products (sensitization)}$$

$${}^3D \longrightarrow \text{products (quenching)}$$

D = donor

A = acceptor

ceptor. Light absorption by the donor produces singlet excited donor ^1D, which undergoes intersystem crossing, giving triplet excited donor ^3D. Triplet excited donor then collides with acceptor, producing triplet excited acceptor ^3A and ground state donor D. The concentration of the acceptor must be kept low enough to make collisions with singlet excited donor improbable. This concentration will be determined by the singlet lifetime of the donor. If ^3A gives the products of interest, this is called a *sensitization mechanism*. If the products of interest are derived from ^3D, A is a *quencher* and this is a *quenching mechanism*. Sensitization and quenching are important methods for determining the spin multiplicity of excited states responsible for photochemical reactions. Sensitization is also an important method for producing the triplet states of molecules in which the efficiency of intersystem crossing is low or zero. The chemistry of singlet and triplet excited states is often quite different.

Let us consider now a specific example, which illustrates the use of sensitization in photochemistry. Direct irradiation of 1,3-butadiene in solution gives cyclobutene and bicyclobutane with minor amounts of dimers. Intersystem crossing efficiency in 1,3-butadiene approaches zero, and triplet-derived products are not formed. Triplet-excited 1,3-butadiene produced by energy transfer from triplet-excited benzophenone gives

cis and
trans

only dimers. Comparison of the ultraviolet absorption spectra of benzophenone and 1,3-butadiene (Figs. 3-1 and 3-2) shows that 3660 Å light will be absorbed only by benzophenone. Benzophenone with its small

S_1–T_1 gap has 100% efficient intersystem crossing. The triplet energy of benzophenone (69 kcal/mole) is quite adequate for diffusion-controlled energy transfer to 1,3-butadiene (E_T = 60 kcal/mole). The energy of the

$$(C_6H_5)_2CO \xrightarrow[\substack{3660\overset{\circ}{A}}]{h\nu} {}^1[(C_6H_5)_2CO]$$

$${}^1[(C_6H_5)_2CO] \longrightarrow {}^3[(C_6H_5)_2CO]$$

$${}^3[(C_6H_5)_2CO] + CH_2 = CH - CH = CH_2 \longrightarrow (C_6H_5)_2CO + {}^3[CH_2 = CH - CH = CH_2]$$

$${}^3[CH_2 = CH - CH = CH_2] + CH_2 = CH - CH = CH_2 \longrightarrow$$

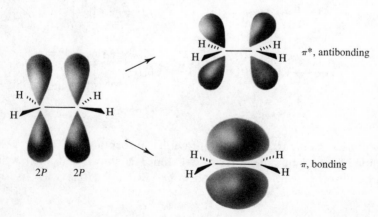

$S_0 \longrightarrow S_1$ transition of benzophenone is lower than that of 1,3-butadiene (see Fig. 3-7). Transfer of singlet energy from excited singlet benzophenone to 1,3-butadiene is thus not expected.

3.8 MOLECULAR ORBITAL VIEW OF EXCITATION

An understanding of electronic excitation in molecular orbital terms is possible only if we consider antibonding as well as bonding orbitals. Imagine an ethylene molecule with the σ framework already formed but with one electron localized on each carbon in a $2p$ atomic orbital. Now let these two electrons interact to form a bond. Just as with a diatomic molecule, two molecular orbitals are formed, a bonding molecular orbital and an antibonding molecular orbital.

Schematically, we can represent the situation as shown below. The two $2p$ atomic orbitals of carbon *mix*, giving rise to a lower energy, bonding molecular orbital (π) and a higher energy, antibonding molecular or-

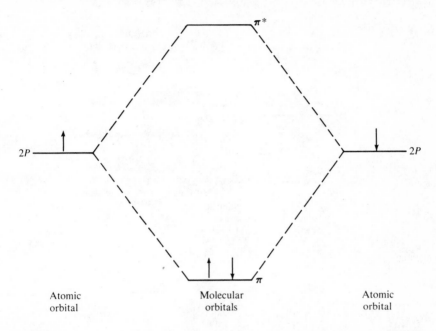

Atomic orbital   Molecular orbitals   Atomic orbital

bital π^*. Each atomic orbital initially contained one electron. In the molecular orbitals both electrons will occupy the lower energy, bonding π molecular orbital. Absorption of a quantum of light excites an electron from the bonding orbital to the antibonding orbital. This is known as a $\pi \longrightarrow \pi^*$ (read "pi to pi-starred") transition. For ethylene this transition occurs at about 1800 Å.

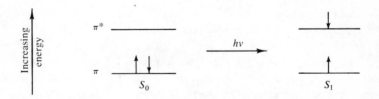

The four π-molecular orbitals of 1,3-butadiene are formed from four $2p$ atomic orbitals (one from each carbon). The lower-energy molecular orbitals that correspond to the wave functions ψ_1 and ψ_2 are bonding orbitals, whereas the higher-energy molecular orbitals corresponding to ψ_3 and ψ_4 are antibonding orbitals. The four π-electrons of 1,3-butadiene occupy the two low-energy bonding orbitals. The lowest-energy electronic transition in 1,3-butadiene is $\pi \longrightarrow \pi^*$ and involves promotion of an electron from ψ_2 to ψ_3.

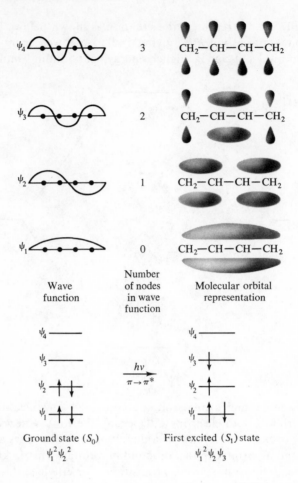

Wave function	Number of nodes in wave function	Molecular orbital representation
ψ_4	3	$CH_2-CH-CH-CH_2$
ψ_3	2	$CH_2-CH-CH-CH_2$
ψ_2	1	$CH_2-CH-CH-CH_2$
ψ_1	0	$CH_2-CH-CH-CH_2$

$$\frac{hv}{\pi \to \pi^*}$$

Ground state (S_0) First excited (S_1) state

$\psi_1^2 \psi_2^2$ $\psi_1^2 \psi_2 \psi_3$

Notice that each wave function ψ_n has $n - 1$ nodes. A node is a point at which the amplitude of the wave function becomes zero as the wave function changes sign. It is a general rule that for linear conjugated polyenes ψ_n will have $n - 1$ nodes. In Chapter 6 we shall make use of the number of nodes of a wave function to predict the stereochemical course of concerted photochemical and thermal reactions.

The carbonyl group presents some new features of interest in terms of a molecular orbital view of excitation. Let us consider a carbonyl group constructed from an sp^2-hybridized carbon and an sp-hybridized oxygen. We shall assume the sigma-bond framework to be present, and we shall direct our attention to the nonbonding orbitals and the π-molecular orbitals. Before formation of the π-molecule orbitals we have $2P_z$ atomic

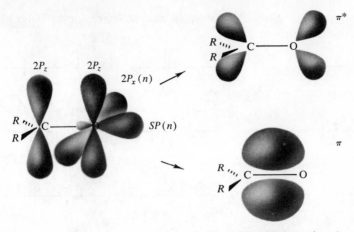

π^*

$2P_z$ $2P_z$

$2P_x(n)$

$SP(n)$

π

(2P_x and SP orbitals not shown)

orbitals on carbon and oxygen. These orbitals are mixed to form a bonding (π) molecular orbital and an antibonding (π^*) molecular orbital similar to those of ethylene. The remaining orbitals of the oxygen atom ($2P_x$ and sp) are doubly occupied and are nonbonding orbitals (n). The $2P_x$ orbital is a relatively high-energy orbital and very important in photochemistry, whereas the sp orbital is a low-energy orbital that is

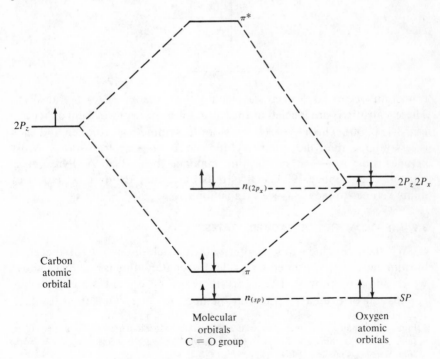

π^*

$2P_z$

$n_{(2p_x)}$

$2P_z\,2P_x$

Carbon
atomic
orbital

π

$n_{(sp)}$

SP

Molecular
orbitals
C = O group

Oxygen
atomic
orbitals

not important in photochemistry. In simple energy level diagrams the $n(sp)$ orbital is ignored.

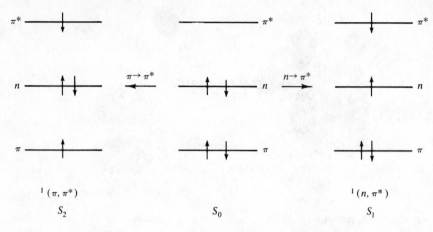

$^1(\pi, \pi^*)$ $^1(n, \pi^*)$

S_2 S_0 S_1

The $n \longrightarrow \pi^*$ transition is the lowest-energy transition for most ketones and is thus the $S_0 \longrightarrow S_1$ transition. The $\pi \longrightarrow \pi^*$ transition corresponds to the $S_0 \longrightarrow S_2$ transition. The lower-energy $n \longrightarrow \pi^*$

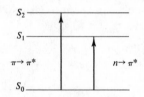

transition occurs at longer wavelength than the $\pi \longrightarrow \pi^*$ transition. These transitions are labeled in the ultraviolet spectrum of benzophenone Fig. 3-1, p. 30). The $n \longrightarrow \pi^*$ transition is symmetry-forbidden and thus is less intense than the symmetry allowed $\pi \longrightarrow \pi^*$ transition.[†] Most ketones show $n \longrightarrow \pi^*$ absorption maxima above 2850 Å. Pyrex glass transmits light above 2900 Å. It is thus quite easy to limit excitation of many ketones to the $n \longrightarrow \pi^*$ transition.

3.9 THE GEOMETRY OF EXCITED STATES

We shall conclude this chapter with a brief discussion of the equilibrium geometry of excited states. Unfortunately, this is a subject about which little is known. The short lifetimes of excited states preclude application of ordinary structural methods to them. One thing is clear,

---

[†] For a discussion of symmetry-allowed and symmetry-forbidden transitions, see H. H. Jaffé and M. Orchin, *Theory and Application of Ultraviolet Spectroscopy* (New York: John Wiley & Sons, Inc., 1969). Chap. 6.

however, and that is that *excited states will often have quite different equilibrium geometries than ground states*. It is known, for example, that excited acetylene is bent rather than linear. Formaldehyde, which is

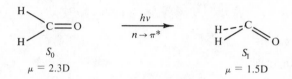

planar in its ground state, distorts slightly toward a pyramidal structure in the S_1 state. The dipole moments of the ground state and first excited

state of formaldehyde are also quite different, indicating significantly different electron distribution.

3.10 PROBLEMS

1. Would you expect the $n \longrightarrow \pi^*$ transition of cyclohexanone to be at longer or shorter wavelength than the $n \longrightarrow \pi^*$ transition of 2-cyclohexenone? Why? What about the $\pi \longrightarrow \pi^*$ transitions?

2. Draw representations of the three wave functions and molecular orbitals for the allyl system. Which orbitals would be occupied in the cation, the anion, and the radical?

REFERENCES

1. H. H. Jaffe and M. Orchin, *Theory and Applications of Ultraviolet Spectroscopy*. New York: Wiley, 1962.

2. M. Orchin and H. H. Jaffe, *The Importance of Antibonding Orbitals*. New York: Houghton Mifflin, 1967.

3. N. J. Turro, *Molecular Photochemistry*. New York: Benjamin, 1965.

4. N. J. Turro, J. C. Dalton, and D. S. Weiss, "*Photosensitization by Energy Transfer*," in *Organic Photochemistry*, Vol. 2, ed. O. L. Chapman. New York: Dekker, 1969, p. 1.

4

Introduction to Photochemical Reactions

4.1 INTRODUCTION

In the last chapter we discussed the nature of electronic excitation, and developed briefly the basic concepts of spin multiplicity and electronic configuration of excited states. We shall now describe some typical photochemical reactions of ketones, olefins, and aromatic compounds.

4.2 REACTIVITY OF ELECTRONICALLY EXCITED KETONES

Ketones have two readily accessible electronic transitions ($n \longrightarrow \pi^*$, $\pi \longrightarrow \pi^*$). In general, the lowest-energy transition is the $n \longrightarrow \pi^*$ transition. This means that the S_1 state of most simple ketones has the n,π^* configuration. In solution, excitation to S_2 will be followed by rapid internal conversion and vibrational equilibration to S_1. The low-lying triplet state T_1 may have either the n,π^* or π,π^* configuration. This comes from the fact that the difference in energy between a $^1(\pi,\pi^*)$ state and the corresponding $^3(\pi,\pi^*)$ state is much larger than the energy difference between a $^1(n,\pi^*)$ state and the corresponding $^3(n,\pi^*)$ state. If the energy difference between the $^1(n,\pi^*)$ and $^1(\pi,\pi^*)$ states is small, it is likely that the T_1 state will be $^3(\pi,\pi^*)$. If the energy difference between the $^1(n,\pi^*)$ and $^1(\pi,\pi^*)$ states is sufficiently large, the T_1 state will be $^3(n,\pi^*)$. The

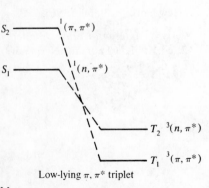

Low-lying π, π^* triplet

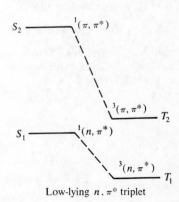

Low-lying n, π^* triplet

reactivity of an excited ketone depends upon the multiplicity of the excited state and upon the electronic configuration of the excited state.
Electronic excited states with the n,π^* configuration show reactivity that
is primarily a result of the singly occupied n-orbital. The vacancy in this
orbital means that n,π^* excited states will undergo reactions that place
an electron in this orbital. Excited states with the π,π^* configuration are
less reactive and longer-lived than n,π^* states.

4.3 REPRESENTATION OF EXCITED STATES OF KETONES

Valence bond representations of excited states of ketones are less
satisfactory than valence bond representations of ketone ground states.
We shall focus attention on the representation of n,π^* states, since these
are the states responsible for much of the interesting photochemistry of
ketones. In an n,π^* state, the presence of an electron in the antibonding
(π^*) orbital reduces the double bond character of the carbon-oxygen
bond, while the singly occupied n-orbital conveys radical-like reactivity to
the oxygen atom. These two ideas have led to representation **A**, which
was popular in the early photochemical literature. This representation
suffers from the implication that the double bond is now a single bond

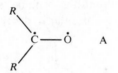

and that free rotation about the carbon-oxygen bond is implied. Furthermore, the geometry of the singly occupied orbital on oxygen is not specified. We shall use representation **B** for ketone n,π^* states but with no
conviction that this is the best representation. The advantages of this

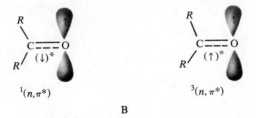

representation are as follows: (1) The vacancy in the $n(2p_x)$ orbital of the
oxygen atom is apparent. (2) The partial double bond is clearly shown.
(3) The spin multiplicity of the state can be shown. (4) The electron in the
π^*-orbital is shown between carbon and oxygen, which is appropriate for
a molecular orbital.

4.4 α-CLEAVAGE

One consequence of the vacancy in the n orbital of an n,π^* excited state of a ketone is the tendency to homolytic cleavage of the α carbon-carbon bond. This process is known as α cleavage or *Norrish type I cleavage*. The half-vacant nonbonding orbital of the n,π^* excited state

$$CH_3-\overset{\overset{\textstyle O}{\|}}{C}-CH_3 \xrightarrow{h\nu} CH_3-\overset{\overset{\textstyle O}{\|}}{C}\cdot + \cdot CH_3$$

overlaps the σ-bond orbital between the carbonyl group and the α-carbon. This overlap facilitates cleavage of the α-bond. The carbon atom of the

$$CH_3-\overset{\overset{\textstyle O}{\|}}{C}\cdot + \cdot CH_3$$

carbonyl group undergoes rehybridization during the process of formation of the acyl radical. Both singlet and triplet n,π^* states undergo α-cleavage.

In molecules with differing degrees of substitution on the α-carbons, α-cleavage will occur in such a way that the most stable alkyl radical is formed. 2,2-Dimethylcyclohexanone, for example, cleaves to give the tertiary alkyl radical rather than the primary alkyl radical. The ultimate products are formed by intramolecular hydrogen transfer reactions.

Suitably placed double bonds or cyclopropane rings greatly facilitate α-cleavage.

The photochemistry of 3,5-cycloheptadienone provides an instructive example. Direct irradiation gives rise to S_1, which has the $^1(n,\pi^*)$ con-

figuration and undergoes α-cleavage, leading ultimately to decarbonylation. The T_1 state produced by energy transfer has the $^3(\pi,\pi^*)$ configuration and leads to isomerization of the diene. In the low-lying singlet-excited state energy is localized primarily on the carbonyl group, whereas in the low-lying triplet-excited state energy is localized mainly on the diene system.

4.5 γ-HYDROGEN TRANSFER

Ketones in n,π^* excited states (either singlet or triplet) frequently undergo hydrogen transfer from a γ-carbon (γ-hydrogen transfer) leading to a 1,4-biradical. Ketones with low-lying π,π^* states do not undergo this

reaction. The 1,4-biradical can revert to starting ketone, close to a substituted cyclobutanol, or cleave to an olefin and the enol of a ketone. The latter process is known as *Norrish type II* cleavage. Solvents that can hydrogen bond to the hydroxyl group of the 1,4-biradical stabilize it and retard the reverse reaction. Replacement of the γ-hydrogen by deuterium leads to formation of a 1,4-biradical with an O-D bond. The energy required to break an O-D bond is larger than the energy required to break an O-H bond, and consequently the reverse reaction is slower and an increase in the efficiency of the overall reaction is observed.

Both $^1(n,\pi^*)$ and $^3(n,\pi^*)$ states give rise to γ-hydrogen transfer. There are, however, distinguishable differences in the singlet and triplet reactions.

S-(+)-5-methyl-2-heptanone

Racemic 5-methyl-2-heptanone

Irradiation of S-(+)-5-methyl-2-heptanone gives initially the $^1(n,\pi^*)$ state, which can intersystem cross to the $^3(n,\pi^*)$ state, return to optically active starting material, or react to give products. The products probably are formed from a singlet biradical. The $^3(n,\pi^*)$ state gives rise to a triplet biradical, which can decay to racemic starting material or go on to products. Racemization of the starting material occurs only by the triplet reaction.

4.6 PHOTOREDUCTION

Photoreduction of ketones is one of the oldest and most thoroughly investigated photochemical processes. In fact, one can carry out the reduction of benzophenone by exposing a solution of benzophenone and benzhydrol in benzene in a Pyrex test tube to ordinary sunlight for a few days. The product, benzpinacol, crystallizes from the solution. The first step in the photochemical reduction is excitation of the benzophenone

$$(C_6H_5)_2CO + (C_6H_5)_2CHOH \xrightarrow[C_6H_6]{hv} \overset{\overset{OH}{|}}{(C_6H_5)_2C}\overset{\overset{OH}{|}}{-C(C_6H_5)_2}$$

benzophenone   benzhydrol   benzpinacol

to the $^1(n,\pi^*)$ state, which intersystem crosses to the $^3(n,\pi^*)$ state. Hydrogen abstraction from benzhydrol gives two diphenylhydroxymethyl radicals, which combine to form benzpinacol. Ketones undergo photore-

$$(C_6H_5)_2C=O \xrightarrow[n\to\pi^*]{hv} {}^1[(C_6H_5)_2C=O]^{n,\pi^*}$$

$$^1[(C_6H_5)_2C=O]^{n,\pi^*} \longrightarrow {}^3[(C_6H_5)_2C=O]^{n,\pi^*}$$

$$^3[(C_6H_5)_2C=O]^{n,\pi^*} + (C_6H_5)_2CHOH \longrightarrow 2\,C_6H_5\overset{\overset{OH}{|}}{\underset{\cdot}{-C}}-C_6H_5$$

$$2\,C_6H_5\overset{\overset{OH}{|}}{\underset{\cdot}{-C}}-C_6H_5 \longrightarrow \overset{\overset{OH}{|}\;\overset{OH}{|}}{(C_6H_5)_2C-C(C_6H_5)}$$

duction in the presence of a variety of hydrogen atom donors other than secondary alcohols. In most cases the mechanisms are similar to that given above with the hydrogen atom donor taking the place of benzhydrol. The effect of substituents in the aryl ring can be quite dramatic because of changes in the electronic configuration of the low-lying triplet state (see Sec. 5.7). Hydrogen atom abstraction is much more efficient for n,π^* states of ketones than for π,π^* states.

Excited ketones can be reduced by amines. The key reaction in this case is electron transfer from the amine to the ketone producing the ketyl

$$^3[C_6H_5COC_6H_5] + R_3N: \longrightarrow C_6H_5\overset{\overset{O^{\ominus}}{|}}{\underset{\cdot}{-C}}-C_6H_5 + R_3N^{\oplus\cdot}$$

radical and the cation radical from the amine.

4.7 PATERNO-BÜCHI REACTION

The Paterno-Büchi reaction involves the addition of an excited ketone to an olefin to form an oxetane. Only ketones with low-lying n,π^* states

$$(C_6H_5)_2C=O \xrightarrow{hv} {}^1(n,\pi^*) \longrightarrow {}^3(n,\pi^*) \xrightarrow{(CH_3)_2C=CH_2} CH_3\overset{\overset{H_2C-O}{|\qquad|}}{\underset{\overset{|\qquad|}{CH_3\;C_6H_5}}{+\qquad+}}C_6H_5$$

2, 2-diphenyl
3, 3-dimethyloxetane

are reactive in this process. Most of the Paterno-Büchi reactions reported involve $^3(n,\pi^*)$ ketones. The major mode of addition can be predicted by assuming that the radical-like oxygen atom of the n,π^* ketone adds to the olefin to give preferentially the most stable biradical intermediate. In the addition of $^3(n,\pi^*)$ benzophenone to trimethylethylene,

$$C_6H_5 - \overset{O}{\underset{}{C}} - C_6H_5 + CH_3CH = C(CH_3)_2 \longrightarrow (C_6H_5)_2C \quad C(CH_3)_2$$

$$O - CHCH_3$$

(n, π^*)

More stable biradical

$$(C_6H_5)_2C\uparrow \quad \uparrow CHCH_3$$
$$O - C(CH_3)_2$$

Less stable biradical

$$(C_6H_5)_2C - C(CH_3)_2$$
$$O - CHCH_3$$

Major product

$$(C_6H_5)_2C - CHCH_3$$
$$O - C(CH_3)_2$$

Minor product

the element of choice lies between a secondary radical and a tertiary radical, since both biradicals are the same in other respects. The tertiary radical is more stable, and this mode of addition is preferred.

The biradical hypothesis is useful in predicting the major product in a Paterno-Büchi reaction, but it is not adequate as a mechanism. The rate constant for the reaction of excited ketone with ground state olefin is very high ($\sim 10^9$ 1 mole^{-1}sec^{-1}) in the few cases that have been studied. This value is several orders of magnitude greater than rate constants for the addition of oxy radicals to olefins. It is probable that the reaction involves an exciplex (complex between excited ketone and olefin) that collapses to the biradical. The biradical is an appealing intermediate,

$$\text{ketone} \xrightarrow{h\nu} {}^1(\text{ketone})$$

$${}^1(\text{ketone}) \longrightarrow {}^3(\text{ketone})$$

$${}^3(\text{ketone}) + \text{olefin} \rightleftarrows \text{exciplex}$$

$$\text{exciplex} \longrightarrow \text{biradical}$$

$$\text{biradical} \longrightarrow \text{oxetane}$$

$$\text{biradical} \longrightarrow \text{ketone} + \text{olefin}$$

because addition of $^3(n,\pi^*)$ benzophenone to *cis* and *trans*-2-butene gives the same mixture of adducts in each case.

Two side reactions can limit the synthetic utility of the Paterno-Büchi reaction. If reactive hydrogen atoms (such as allylic hydrogen atoms) are present in the olefin, hydrogen abstraction by the excited ketone will compete with the Paterno-Büchi reaction, and complex mixtures will be formed. If the triplet energy of the ketone is comparable to, or exceeds, that of the olefin, energy transfer will compete with or supplant addition. The problem is especially keen with aliphatic ketones because of their high triplet energies. Acetone ($E_T \sim 78$ kcal/mole), for example, transfers triplet energy to norbornene and thus produces dimers, whereas benzophenone ($E_T = 69$ kcal/mole) adds to norbornene.

Acetylenes can replace olefins in the Paterno-Büchi reaction. The initial adducts are assumed to be oxetenes, but the only products isolated are α,β-unsaturated ketones.

$$(C_6H_5)_2CO + C_4H_9C \equiv C\,C_4H_9 \xrightarrow{h\nu}$$

Reactions of $^1(n,\pi^*)$ excited ketones with electronegatively substituted olefins are known which also give oxetanes. These reactions differ from

$$R_2CO \xrightarrow{h\nu} {}^1(R_2CO)$$

the usual Paterno-Büchi reaction in substituent effects on the olefin and in retention of the stereochemical integrity of the olefin. The $^1(n,\pi^*)$ excited states thus seem to be nucleophilic, and the addition probably does not involve biradical intermediates.

4.8 REACTIVITY OF π,π^* EXCITED KETONES

Ketones with low-lying π,π^* states are not reactive in α-cleavage, hydrogen abstraction, or the Paterno-Büchi reaction. These reactions depend on the vacancy in the nonbonding orbital of oxygen characteristic of n,π^* states. Reduction of π,π^* excited ketones by electron transfer from amines has been observed. The best-known reactions of π,π^* excited ketones are cycloadditions and dimerizations of unsaturated ketones.

Triplet π,π^* states have longer lifetimes than triplet n,π^* states (see Sec. 5.7.). Self-quenching and excimer (excited dimer) formation are more common for $^3(\pi,\pi^*)$ states than for $^3(n,\pi^*)$ states. Excimer formation

$$R_2CO + R_2CO \longrightarrow 2\,R_2CO \quad \text{self-quenching}$$
$$^3(\pi,\pi^*)$$

$$R_2CO + R_2CO \rightleftharpoons \text{excimer} \longrightarrow 2\,R_2CO$$
$$^3(\pi,\pi^*) \qquad\qquad\qquad\qquad \downarrow$$
$$\text{dimers}$$

can lead to dimers or to self-quenching. Self-quenching can cause difficulty in the use of $^3(\pi,\pi^*)$ sensitizers, because it competes with the desired energy transfer process. Rate constants for self-quenching can be quite large (10^8–10^9 l mole^{-1}sec^{-1}).

4.9 PHOTOCHEMISTRY OF α,β-UNSATURATED KETONES

The photochemistry of α,β-unsaturated ketones is rich and varied with many facets of mechanistic interest. We can describe only a few of the more representative processes.

Irradiation of 2-cyclohexenone gives rise to two major photodimers. These dimers may be formed via an excimer (excited dimer) derived from the $^3(\pi,\pi^*)$ cyclohexenone and a molecule of ground state cyclohexenone. The ratio of the two dimers depends on the solvent used. Polar solvents

head-to-head   +   head-to-tail

favor formation of the head-to-head dimer. A related reaction can be carried out using olefins in place of the second cyclohexenone. Irradiation of cyclohexenone in the presence of 1,1-dimethoxyethylene, for example, gives two adducts. The highly strained *trans* adduct is the major product.

$$+ (CH_3O)_2C = CH_2 \xrightarrow{\;hv\;}$$

cis
21%

+

trans
49%

When the enone system is in a seven- or eight-membered ring, *cis-trans* isomerism rather than dimer formation occurs. These *trans*-enones are reactive and unstable, but they can be studied by spectroscopic methods. *trans*-2-Cyclooctenone has been isolated.

Conjugated ketones undergo a variety of complex photochemical rearrangements, some of which have been studied in considerable detail. Among the earliest system to be investigated was the terpene *santonin*. The photochemical transformations of santonin give insight into both the complexities and the power of the photochemical method. We shall consider this system in some detail in order to illustrate what can happen in a moderately complex molecule.

Irradiation of santonin in refluxing aqueous acetic acid gives an ester of *isophotosantonic lactone*, with an empirical formula that differs from that of santonin by the addition of a molecule of water, whereas irradiation in ethanol gives an isomer of santonin, *lumisantonin*. In both of these products, extensive skeletal rearrangement has taken place. Lumisantonin undergoes photochemical rearrangement to a ketene that cyclizes thermally to *mazdasantonin* or reacts with water or alcohol to give *photosantonic acid* or its ester. Irradiation of mazdasantonin also gives the ketene. This process is an example of α-cleavage in an unsaturated ketone.

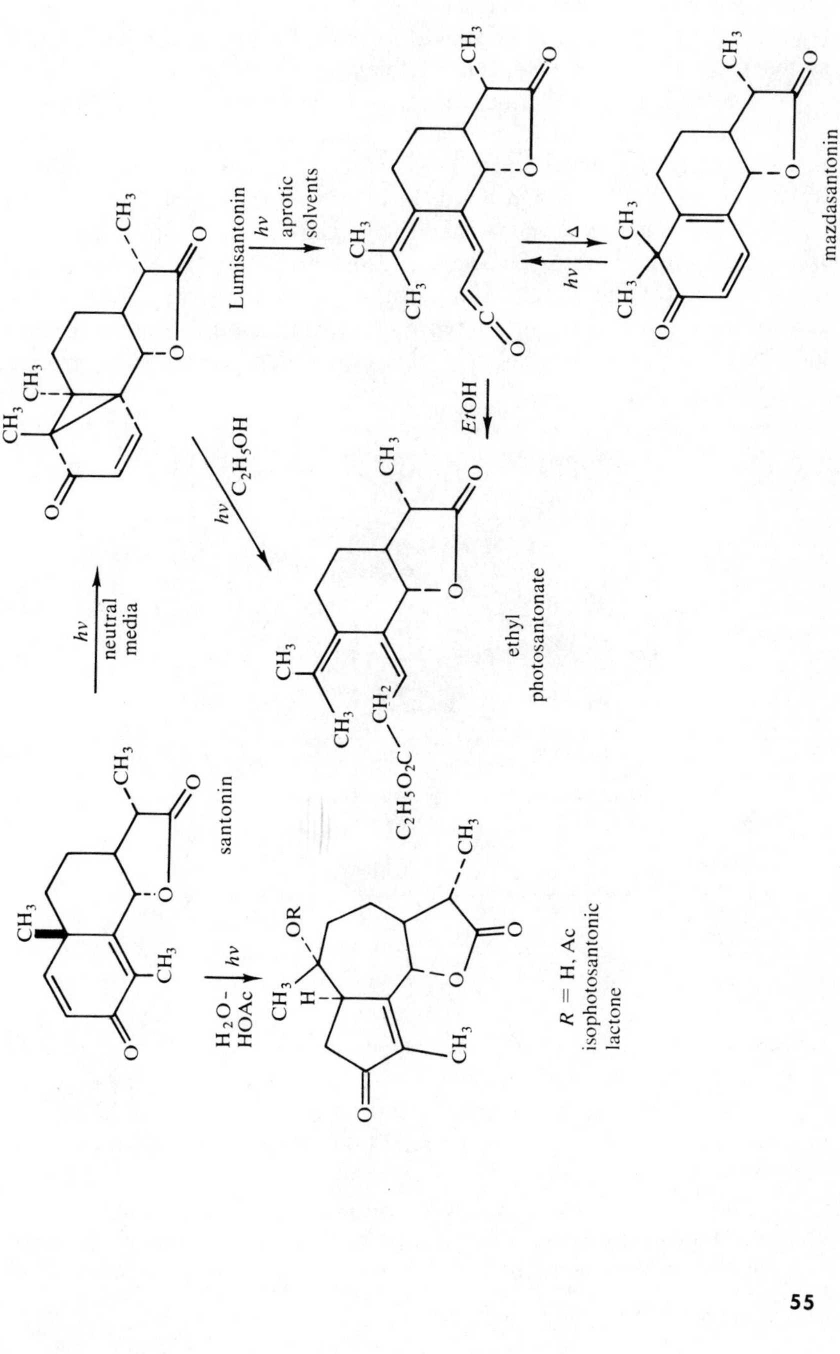

Lumisantonin

hv aprotic solvents

Δ / *hv*

mazdasantonin

*Et*OH

ethyl photosantonate

hv neutral media

hv C₂H₅OH

santonin

H₂O– HOAc | *hv*

R = H, Ac isophotosantonic lactone

55

4.10 OLEFIN PHOTOCHEMISTRY

Having no *n* electrons, alkenes have no n,π^* excited states. Thus their photochemistry involves only π,π^* states and affords a contrast to the photochemistry of ketones. We can also examine more closely the chemistry of singlet excited states as opposed to that of triplet states, since intersystem crossing from S_1 to T_1 is negligible in many olefins.

The most common photochemical transformation of the T_1 state of simple olefins is *cis-trans* isomerization. Since the π^* orbital is antibonding, the energy of the system can be reduced by twisting approximately 90° about the carbon-carbon bond. In this form, the distinction between *cis* and *trans* is lost, and the twisted triplet can return to either the *cis* or *trans* olefin ground state. Twisting is also favorable in S_1. A calculated energy curve for the S_0, S_1, and T_1 states is given in Fig. 4-1. In the ground state, the 90° orientation represents a maximum in energy because all π bonding is lost, whereas in T_1 this orientation represents an energy minimum.

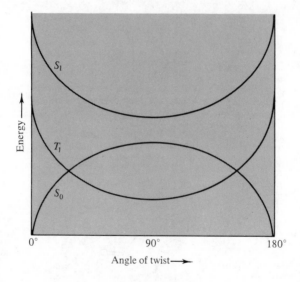

Fig. 4-1

Simple nonconjugated olefins have high triplet energies (>75 kcal/mole), and sensitizers with equally high triplet energies are required for energy transfer. Benzene, toluene, xylene, or acetone are suitable for this purpose. Toluene and xylene are usually preferable to benzene because of their greater photostability.

Triplet states of cyclohexenes and cycloheptenes, which cannot undergo facile *cis-trans* isomerization, protonate readily. "Acids" as weak as methanol are sufficient, and a sensitized irradiation of a cyclic olefin in methanol makes a useful synthetic method for producing an ether. The intermediacy of a carbonium ion is assumed, because typical carbonium

ion rearrangements have been observed in suitable systems.

Olefins may also undergo photochemical dimerization, just as unsaturated ketones do. The dimers are formed by reaction of triplet olefin with ground state olefin. This reaction will be treated in more detail in Chapter 6. Intramolecular as well as intermolecular examples of this reaction are known. The photoisomerization of norbornadiene to quadricyclene occurs on direct or sensitized irradiation.

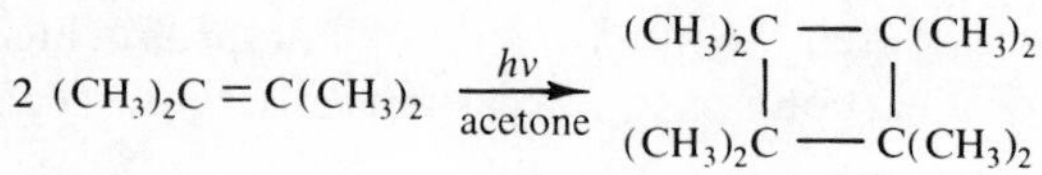
2 (CH3)2C = C(CH3)2
hv
acetone
(CH3)2C — C(CH3)2
(CH3)2C — C(CH3)2

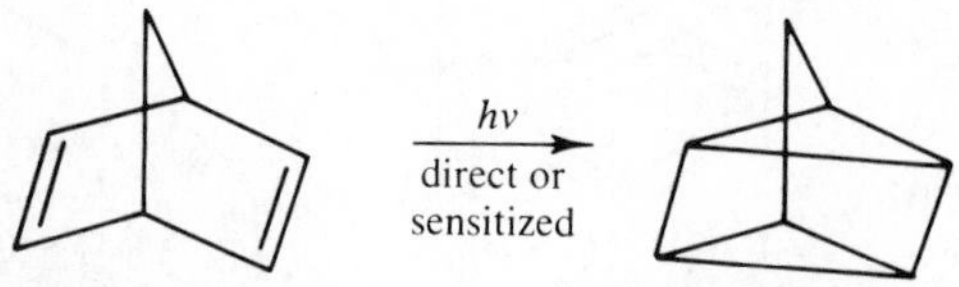
hv
direct or
sensitized

norbornadiene quadricyclene

In 1,5-hexadiene systems that have greater flexibility than norbornadiene, mercury-sensitized addition of one double bond to the other usually occurs in a "crossed" sense, giving rise to bicyclic systems. Although it has not been established for certain whether or not the reaction occurs by way of a discrete biradical intermediate, this example is written with

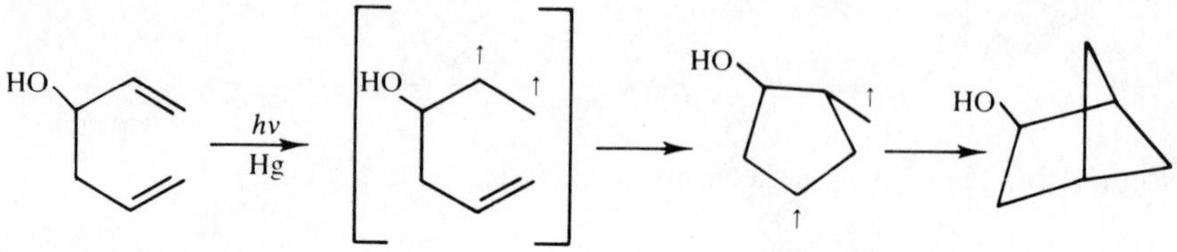
HO
hv
Hg
HO
HO
HO

such an intermediate so that the formation of the product may be more easily visualized.

Cyclopropane rings are also formed photochemically from many simple olefins. The cyclopropanes arise by 1,2-hydrogen shifts followed by cyclization. 1,5-Cyclooctadiene affords an example in which "crossed" addition and cyclopropane formation are both observed.

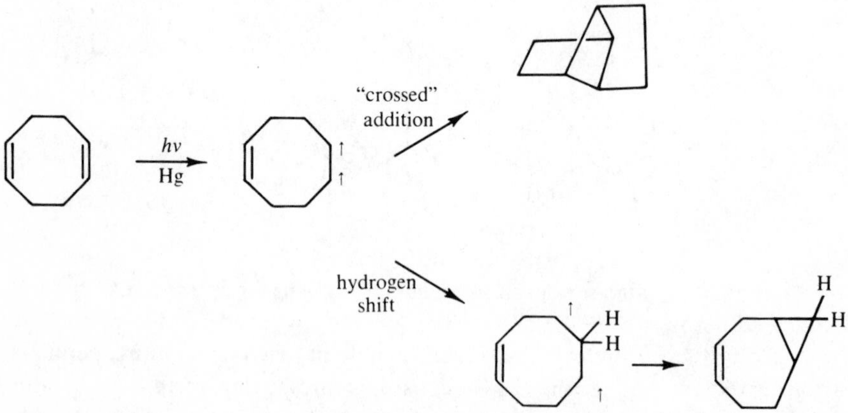
hv
Hg
"crossed"
addition
hydrogen
shift
H
H
H
H

An intriguing and general rearrangement of divinylmethane derivatives to vinylcyclopropanes has been observed. Bicyclo-[2.2.2]octa-2,5,7-

triene (barrelene), for example, on irradiation in the presence of a sensitizer (acetone) gives semibullvalene and cyclooctatetraene. Semibullvalene undergoes a sensitized photoisomerization to cyclooctatetraene.

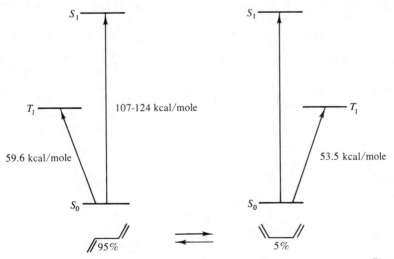

Bicyclo [2.2.2] octa-2,5,7-triene
(Barrelene)   Semibullvalene

4.11 CONJUGATED OLEFINS

The photochemistry of conjugated olefins, like that of conjugated ketones, is more complex than that of the simple analogs. The fascinating and vital photochemical reactions associated with ergosterol and vitamin D will be dealt with in Chapter 6. For the present, we shall consider some of the simpler aspects of the photochemistry of conjugated dienes.

1,3-Butadiene exists in solution as a rapidly equilibrating mixture of *transoid* and *cisoid* conformers, with the former predominating by nearly 20:1. Since light absorption occurs without any change in nuclear positions, there are excited states corresponding to each rotational isomer. In Figure 4-2, the energetics of the system are shown. The exact energies of the S_1 states are not known, but cisoid S_1 probably lies below transoid S_1. The large energy gap between S_1 and T_1 accounts for the fact that intersystem crossing does not occur. Direct irradiation of 1,3-butadiene in solution thus gives rise to chemistry only from the S_1 state.

S_1 —————

S_1 —————

T_1 ————   107-124 kcal/mole   ———— T_1

59.6 kcal/mole   53.5 kcal/mole

S_0 ———— S_0 ————

95%   ⇌   5%

Fig. 4-2

The products of the irradiation of 1,3-butadiene in solution are cyclo-butene and bicyclobutane, in a proportion depending upon the solvent used. It is important to reemphasize that the reaction products discussed in this chapter are found in *solution* photochemistry. Neither cyclobutene

$$CH_2{=}CH{-}CH{=}CH_2 \quad \xrightarrow{\;h\nu\;} \quad \begin{array}{l} CH_2{-}CH \\ \;|\qquad\; || \\ CH_2{-}CH \end{array} \qquad \begin{array}{c} CH \\ CH_2 \diagup\; |\; \diagdown CH_2 \\ \diagdown CH \diagup \end{array}$$

nor bicyclobutane is found in the vapor phase irradiation of 1,3-butadiene.

Irradiation of 1,3-dienes in which the diene conformation is deter-mined by the structure suggests that *cisoid*-1,3-butadiene may be the pre-cursor of cyclobutene and that *transoid*-1,3-butadiene may be the precur-sor of bicyclobutane.

If a concentrated solution of 1,3-butadiene is irradiated, some dimer formation can be detected, although even under the best of circumstances less than 10% dimer formation has been reported. We expect dimer forma-tion to be concentration dependent, since it involves the *bimolecular* reaction of an excited butadiene with a ground state butadiene. The ring-closure reactions, on the other hand, are unimolecular. Only when the butadiene concentration is high can dimerization compete. The relative amounts of the four dimers formed in the direct irradiation are shown

$$^1\!\left[CH_2{=}CH{-}CH{=}CH_2\right] \;+\; CH_2{=}CH{-}CH{=}CH_2 \longrightarrow$$

30%

50%   8%

above for comparison with the products from triplet dimerization.

The chemistry of the T_1 state of 1,3-butadiene is quite different from that of the S_1 state. A mixture of dimers is formed in greater than 75% yield, and neither cyclobutene nor bicyclobutane can be detected in the sensitized (triplet) reaction. The dimer mixture obtained from T_1 also differs significantly from that given by S_1. The dimer composition also depends upon the energy of the sensitizer used to populate T_1. The results for the two sensitizers, acetophenone and benzil, are given in Table 4.1.

Table 4.1

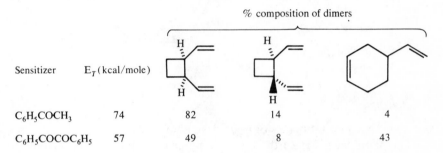

Sensitizer	E_T (kcal/mole)	% composition of dimers		
$C_6H_5COCH_3$	74	82	14	4
$C_6H_5COCOC_6H_5$	57	49	8	43

The origin of this remarkable change in product composition lies in the relative energies of the sensitizers. Acetophenone, with E_T = 74 kcal/mole, is sufficiently energetic to transfer energy to either *cisoid* or *transoid* butadiene. Since the transoid form predominates by a large margin, and since energy transfer occurs at nearly every collision, the dimer composition from the acetophenone-sensitized experiment must represent primarily the reaction of *transoid*-T_1. Benzil (E_T = 54 kcal/mole), on the other hand, is energetic enough to transfer energy to *cisoid* butadiene but not *transoid* butadiene. The 49:8:43 ratio of dimers must represent the dimer production from *cisoid*-T_1. If a sensitizer with a triplet energy of less than 54 kcal/mole were used, no dimer formation would be observed.

A more detailed analysis of the product ratio tends to confirm this analysis. We hypothesized that when acetophenone is the sensitizer, *transoid*-T_1 is the reacting species, adding to ground state butadiene. Since the latter is over 95% *transoid* itself, we have a *transoid* triplet to *transoid* ground state addition. The geometry of this intermediate insures that only *cis* and *trans* divinylcyclobutanes will be formed, since allyl

transoid-T_1

cis and *trans*

radicals are known to retain their geometric identity.

In the benzil-sensitized reaction, *cisoid-T₁* adds to *transoid* ground state butadiene. The geometry of this intermediate is such that it can close

cisoid-T₁   *cis* and *trans*

either to divinylcyclobutanes or 4-vinylcyclohexene.

The photochemistry of the terpene, myrcene, provides an example in which monomeric products derived from the excited singlet and excited triplet states are quite different. Direct irradiation of myrcene gives a complex mixture of products, two of which are shown. The triplet state of myrcene, formed by energy transfer from triplet benzophenone, gives only the bicyclo[2.1.1]hexane derivative shown.

The photochemistry of benzobarrelene provides another example of the differences that can be observed in the chemistry of singlet and triplet excited states. In this case the singlet reaction involves benzovinyl bridg-

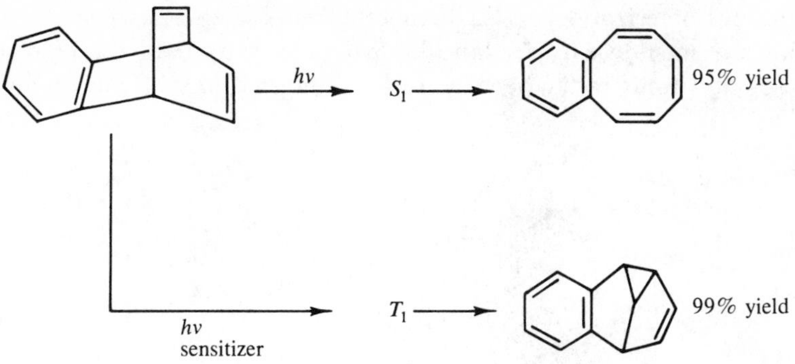

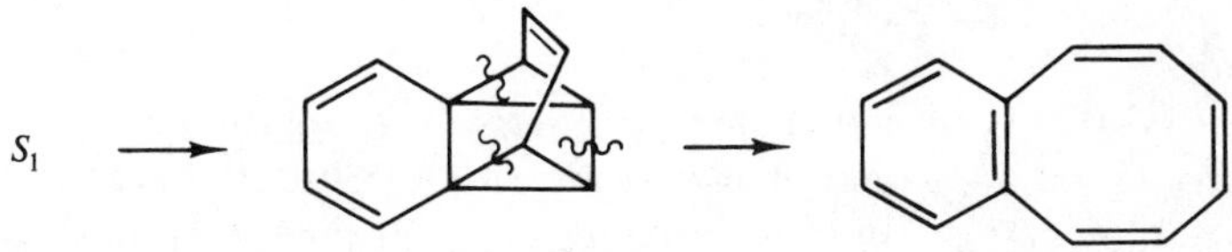

95% yield

99% yield

$S_1 \longrightarrow$

ing giving an intermediate that cleaves three sigma bonds, giving benzo-cyclooctatetraene. The triplet reaction involves vinyl-vinyl bridging.

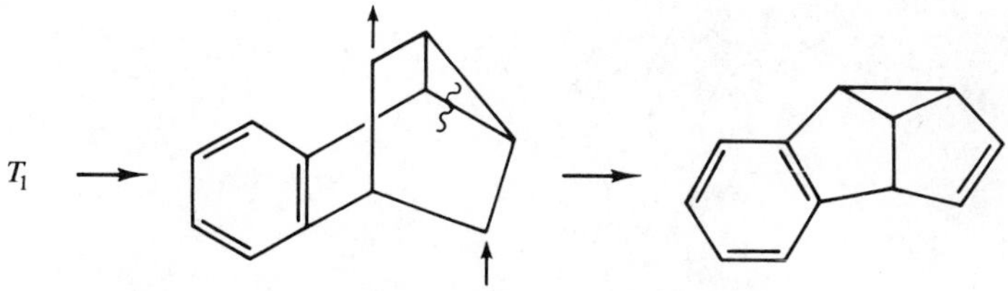

$T_1 \longrightarrow$

4.12 PHOTOCHEMISTRY OF AROMATIC COMPOUNDS

Aromatic compounds undergo many photochemical transformations. The photochemical transformations of benzene derivatives provide a surprising contrast to the thermal stability of aromatic systems. The photochemical rearrangements of simple benzene derivatives provide direct routes to several highly strained molecules. Irradiation of benzene, for example, gives benzvalene and fulvene. Furthermore, alkyl benzenes undergo photochemical isomerization to other isomeric alkyl benzenes.

benzvalene   fulvene

The photoisomerization of 1,3,5-trimethylbenzene to 1,2,4-trimethylbenzene has been shown by ^{14}C-labeling experiments to involve a rearrangement of the atoms in the benzene ring. Additional insight concerning the

$$\blacksquare = {}^{14}C$$

complexity of this process comes from the investigation of the photochemistry of 1,3,5-tri-*t*-butylbenzene and 1,2,4-tri-*t*-butylbenzene. In this system a photostationary state (a pseudo-equilibrium) is established which involves a benzvalene derivative, a prismane derivative, and a Dewar benzene derivative, as well as the two tri-*t*-butylbenzenes. The figures in parentheses give the composition of the photostationary state mixture.

prismane   Dewar benzene

Benzene derivatives also add to olefins and dienes. The addition to simple olefins involves rearrangement as well as addition. In the addition to butadiene there is reason to believe that the double bond in that portion of the molecule derived from butadiene is *trans*. Naphthalenes, anthracenes, and polyacenes in general undergo photodimerization. Anthracene may be taken as an example. Two anthracene molecules join

anthracene

at the 9,10-positions in the dimerization. Substituents at the 9-position give rise to head-to-tail dimers. In some anthracenes the nature of the photochemistry depends on the wavelength of light. Long wavelengths favor photodimerization, and shorter wavelengths favor other processes. The wavelength dependence is due, at least in part, to the photochemical cleavage of the dimers by the shorter wavelength light. Irradiation of 9-nitroanthracene with long wavelength light gives the dimer, whereas shorter wavelength light gives nitric oxide ($\cdot$NO), anthraquinone, anthra-

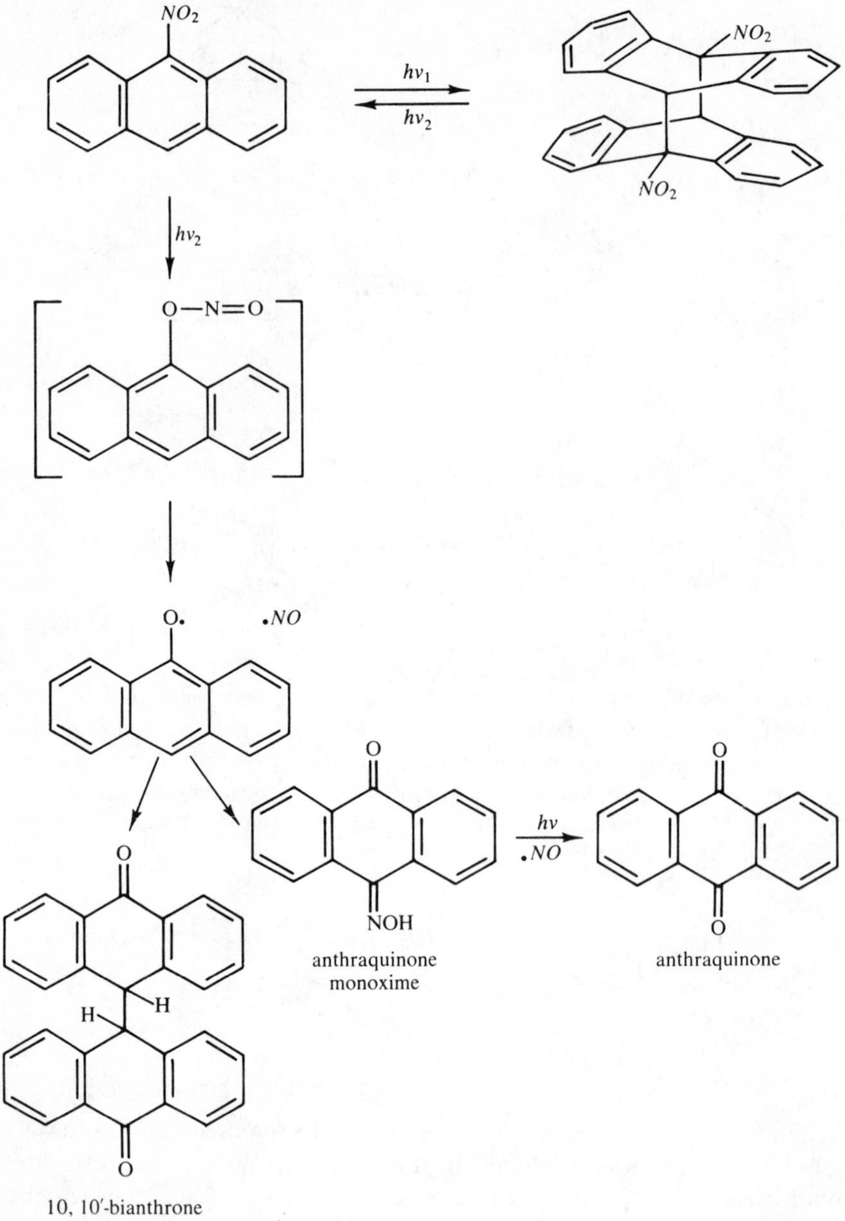

anthraquinone
monoxime

anthraquinone

10, 10'-bianthrone

quinone monoxime, and 10,10'-bianthrone. The latter products presumably are formed via 9-anthryl nitrite.

The electron distribution in an excited state is often very different from that in the ground state. This can be illustrated dramatically in the aromatic series. One example is provided by comparison of the dissociation constant of a phenol in its excited state with that of its ground state. This difference can be measured relatively easily in some cases, since fluorescence from a phenol occurs at a different wavelength from that of the corresponding phenolate ion. If light is absorbed by a phenol, and if the excited state is a strong acid, then dissociation will occur and phenolate fluorescence will occur. By appropriate adjustment of the pH of the solution the dissociation constant of the excited state can be determined. Results for 2-naphthol and 2-naphthylamine are given in Table 4.2. Notice the very large charges in pK_a between S_0 and S_1, indicative of major changes in electron density. The pK_a's of the T_1 states are somewhat closer to those of the ground state.

Table 4.2

Compound	pK$_a$		
	S_0	S_1	T_1
2-naphthol	9.5	3.1	8.1
2-naphthylamine	4.1	−2	3.3

4.13 PROBLEMS

1. Write a mechanism for the reaction shown below.

$$(CH_3)_2C\!=\!C(CH_3)_2 \xrightarrow[\text{acetone}]{\substack{h\nu \\ 3000\text{Å}}} \begin{array}{c} (CH_3)_2C\!-\!C(CH_3)_2 \\ | \quad\quad | \\ (CH_3)_2C\!-\!C(CH_3)_2 \end{array}$$

2. Predict the structure of the major photoadduct in each of the following reactions.

(a) $(C_6H_5)_2CO + CH_2\!=\!CHOCH_3 \xrightarrow{3600\text{Å}}$

(b) $C_6H_5CHO + CH_3CH\!=\!C(CH_3)_2 \xrightarrow{3000\text{Å}}$

(c) $(C_6H_5)_2CO + CH_3C\!\equiv\!CCH_3 \xrightarrow{3660\text{Å}}$

3. Write a mechanism for the photochemical rearrangement below.

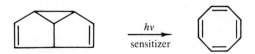

REFERENCES

1. H. E. Zimmerman, *Angewandte Chemie International Edition* (English), **8** (1969), 1.

2. D. C. Neckers, *Mechanistic Organic Photochemistry*. New York: Reinhold, 1967.

3. A. W. Noyes, Jr., G. S. Hammond, and J. N. Pitts, *Advances in Photochemistry*, Vols. 1–6. New York: Interscience, 1967.

4. O. L. Chapman, Ed., *Organic Photochemistry*, Vols. 1, 2. New York: Dekker, 1967.

5
Study
of the Mechanisms
# of Photochemical
Reactions

5.1 INTRODUCTION

In any discussion of reaction mechanism it is important to keep in mind the difference between the study of structure and the study of mechanism. We can prove that a given compound has a certain structure, but we cannot prove that a reaction proceeds by a certain mechanism. We can gather facts that greatly limit the kinds of mechanisms that may be considered for a given reaction. We may, in fact, be able to exclude all but one of the mechanisms that we can imagine. It would be more than audacious, however, to say that we have imagined all possible mechanisms. New observations and greater imagination constantly generate new mechanisms. What we seek in mechanism studies is an understanding of a reaction which is based on adequate observations, which accounts in a logical manner for these observations, and which makes useful predictions about how the reaction will respond to changes in substrate or conditions. Reaction mechanisms that fulfill these requirements are said to be generally accepted mechanisms. All reaction mechanisms need to be examined continually as new observations become available to see if they account adequately for these observations. The study of photochemical mechanisms makes use of many of the techniques common in the study of thermal chemical mechanisms such as labeling experiments, stereochemistry, kinetics, absorption spectroscopy, and trapping of intermediates. In addition, studies of photochemical mechanism make use of specialized techniques such as the determination of quantum efficiencies, sensitization, quenching, and emission spectroscopy. Any study of mechanism must be based on good structure work. A precise knowledge of the structure of starting materials and products is the foundation for sound deductions concerning mechanism.

5.2 DETECTION OF INTERMEDIATES

The photochemical conversion of α-tropolone methyl ether to methyl 4-oxo-2-cyclopentenyl acetate in aqueous solution is a rather complex process. It provides a good example of the importance of detecting intermediates and identifying the individual reactions that comprise the overall transformation observed. Water is involved in the reaction, so the first obvious change in conditions in a search for intermediates is to run

α-tropolone
methyl ether

methyl 4-oxo-2-cyclopentenyl
acetate

the reaction in an aprotic solvent. Under these conditions a new product, 7-methoxybicyclo[3.2.0]hepta-3,6-dien-2-one, is isolated. Irradiation of this product in water gives methyl 4-oxo-2-cyclopentenyl acetate. This experiment does not prove that 7-methoxybicyclo[3.2.0]hepta-3,6-dien-2-one is an intermediate in the reaction, but it does show that it might be. When the irradiation of α-tropolone methyl ether in an aprotic solvent is monitored by vapor chromatography, it is found that the disappearance of starting material correlates with the appearance of a new product, 1-methoxybicyclo[3.2.0]hepta-3,6-dien-2-one. The concentration of the new product rises to a maximum and then decreases with concurrent formation of 7-methoxybicyclo[3.2.0]hepta-3,6-dien-2-one (Fig. 5-1). A summary of these photochemical processes is now possible. The first process is an electrocyclic reaction of the type described in Chapter 6.

α-tropolone
methyl ether

1-methoxybicyclo-
[3.2.0] hepta-3,
6-dien-2-one

7-methoxybicyclo-
[3.2.0] hepta-3,
6-dien-2-one

methyl 4-oxo-2-cyclo-
pentenyl acetic acid

The final process may be viewed as a photochemical (or thermal) hydration of the cyclobutene double bond followed by a reverse Aldol reaction. The process of greatest interest is the photochemical isomerization of one

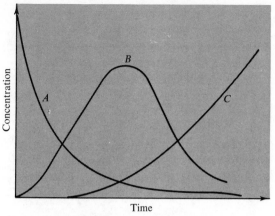

A: α-tropolone methyl ether
B: 1-methoxybicyclo [3.2.0] hepta-3, 6-dien-2-one
C: 7-methoxybicyclo [3.2.0] hepta-3, 6-dien-2-one

Fig. 5-1

bicyclic system to another. This process merits further examination. On the surface, it appears that the methoxyl group moves from carbon 1 to

carbon 7. This, however, is not the case. The movement of atoms that occurs is more complex than a simple shift of the methoxyl group. The nature of the shifts involved was delineated by a series of labeling experiments in which alkyl groups were used as labels. The individual rearrangements studied are shown below.

The implication of these studies is that carbon atoms 1 and 7 and carbon atoms 5 and 6 exchange places in the rearrangement, while carbon atom 4 stays in place. A different kind of experiment shows that the rearrange-

■, ◆, ●, and ▲ designate labels at these positions

ment is even more complex. Irradiation of 1-methoxybicyclo[3.2.0]hepta-3,6-dien-2-one at liquid nitrogen temperature (77°K, −196°C) shows the presence of an intermediate that has a ketene carbonyl band in the infra-red spectrum. This ketene is also formed on irradiation of 7-methoxybi-cyclo[3.2.0]hepta-3,6-dien-2-one. The ketene isomerizes thermally at −70°C to the bicyclic ketones. The ketene has been assigned the structure cis,cis-2-methoxybicyclo[2.1.0]pent-2-en-5-yl ketene. The photochemical

cis, cis-2-methoxybicyclo-
[2.1.0] pent-2-en-5-yl ketene

conversion of the bicyclic ketones to the ketene is an α-cleavage (Norrish type I) process which probably occurs in a $^1(n,\pi^*)$ excited state (see Chapter 4). The thermal isomerization of the ketene to the bicyclic ketones is a Cope rearrangement (Chapters 2 and 6). This rearrangement occurs in a cis-divinylcyclopropane system (see Chapter 2). The presence of the reactive ketene moiety and the high ground state energy of the bicyclopentene system make the rearrangement even more facile. Rearrangement through the ketene accounts for the labeling experiments.

■, ◆, ●, ▲ designate labels at these positions

Sensitization studies have shown that the path from the 1-methoxy-bicyclo[3.2.0]hepta-3,6-dien-2-one to the ketene involves the S_1 state $^1(n,\pi^*)$ of the ketone while the T_1 state gives a different product (6-methoxybicyclo[3.2.0]hepta-3,6-dien-2-one), probably via a biradical intermediate.

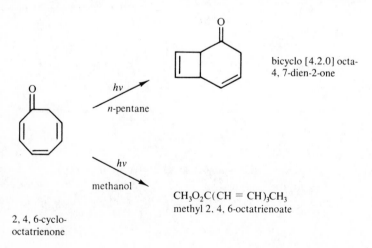

5.3 LOW-TEMPERATURE PHOTOCHEMISTRY

In the last section we encountered the problem of an unstable ground state intermediate (the ketene) in a photochemical process. One of the techniques that is especially useful for detecting unstable intermediates is irradiation at sufficiently low temperature so that the intermediate has a reasonable lifetime. Various spectroscopic techniques are used to detect the presence of the intermediate and to obtain some idea of the chemical nature of the intermediate. Chemical methods are then used to trap the intermediate for detailed structure analysis. We shall now examine an example of the use of low-temperature techniques.

Irradiation of 2,4,6-cyclooctatrienone in *n*-pentane gives a bicyclic photoisomer, whereas irradiation in methanol gives geometric isomers of methyl 2,4,6-octatrienoate. Formation of the ester in methanol clearly suggests the possibility of a ketene intermediate, but such an intermediate

bicyclo [4.2.0] octa-
4, 7-dien-2-one

hv
n-pentane

hv
methanol

$CH_3O_2C(CH = CH)_3CH_3$
methyl 2, 4, 6-octatrienoate

2, 4, 6-cyclo-
octatrienone

had not been observed. There is no *a priori* reason to expect an intermediate in the formation of the bicyclic photoisomer, because this is an allowed electrocyclic reaction (Chapter 6).

Low-temperature irradiations are carried out in an infrared cell enclosed in a vacuum shroud with sodium chloride windows. After irradiation, the cell is transferred to an infrared spectrometer to scan the spectrum. Irradiation of 2,4,6-cyclooctatrienone at 77°K (liquid nitrogen cooling) leads rapidly to several new bands in the infrared spectrum. The carbonyl region of the infrared spectrum shows new bands at 2115 cm^{-1} and 1731 cm^{-1}. No band at 1710 cm^{-1} (carbonyl frequency of bicyclo-[4.2.0]octa-4,7-dien-2-one) is observed. This compound is thus not a primary product under these conditions. The absorption band at 2115 cm^{-1} is due to the ketene. Subsequent experiments have identified the 1731 cm^{-1} species as *trans,cis,cis*-2,4,6-cyclooctatrienone. This highly strained compound can be trapped as the Diels-Alder adducts with furan, and it is the precursor of bicyclo[4.2.0]octa-4,7-dien-2-one.

trans, cis, cis-
2, 4, 6-cyclo-
octatrienone

CH_3OH

$CH_3O_2C(CH = CH)_3CH_3$

5.4 QUANTUM EFFICIENCY

We now need to introduce a new term, quantum efficiency, which occupies a prominent place in studies of photochemical mechanisms. Quantum efficiency (or quantum yield) is a measure of the efficiency of the use of light in a photochemical reaction. The quantum efficiency for the formation of a product (Φ_{form}) is defined as the number of molecules of product formed per quantum of light absorbed. The quantum efficiency

$$\Phi_{form} = \frac{\text{number of molecules of product formed}}{\text{number of quanta absorbed}}$$

$$\Phi_{dis} = \frac{\text{number of molecules of starting material destroyed}}{\text{number of quanta absorbed}}$$

for disappearance of starting material (Φ_{dis}) is defined as the number of molecules of starting material that disappear per quantum of light absorbed. If a photochemical reaction gives only one product, the quantum efficiency of formation and the quantum efficiency of disappearance will be equal. If several products are formed, the sum of the quantum efficiencies of formation will be equal to the quantum efficiency of disappearance. Quantum efficiencies for reactions that do not go through chain mechanisms have values of zero to one. Free radical chain processes have quantum efficiencies as high as several thousand. These high efficiencies arise from the fact that the photochemical reaction is the initiation step in a free radical chain process. The free radical chlorination of methane, for example, has a high quantum efficiency, because the chain length is long. The chain propagation steps occur many times before the chain terminates.

$$Cl_2 + h\nu \longrightarrow 2\,Cl\cdot \quad \} \quad \text{initiation}$$

$$Cl\cdot + CH_4 \longrightarrow CH_3\cdot + HCl$$
$$CH_3\cdot + Cl_2 \longrightarrow CH_3Cl + Cl\cdot \quad \Big\} \quad \text{propagation}$$

$$CH_3\cdot + Cl\cdot \longrightarrow CH_3Cl$$
$$2\,CH_3\cdot \longrightarrow CH_3{-}CH_3 \quad \Big\} \quad \text{termination}$$
$$2\,Cl\cdot \longrightarrow Cl_2$$

5.5 SENSITIZATION

Sensitization experiments play a prominent role in the study of photochemical mechanisms. The basic aspects of sensitization have been covered in Chapter 3. Sensitization studies are often used to show that a triplet excited state is (or is not) involved in a photochemical process. For a satisfactory sensitization experiment in which triplet energy is transferred, a sensitizer of known triplet energy, intersystem crossing efficiency (Φ_{ic}), and intersystem crossing rate constant (k_{ic}) should be used. It has been assumed in many studies in the past that high intersystem crossing efficiencies were an adequate guarantee of high intersystem crossing rate. This is not so. Triphenylene, for example, which has a high intersystem crossing efficiency, has a relatively long-lived singlet excited state (which requires a relatively small rate constant for inter-system crossing).

Sensitizer	E_T(kcal/mole)	Φ_{ic}	k_{ic}(sec^{-1})
Acetophenone	74	1.0	10^{10}
Benzophenone	68	1.0	10^{11}
Triphenylene	68	0.95	$\sim 10^8$

Singlet energy transfer can be minimized by selecting sensitizers that have a singlet transition ($E_{s_0 \to s_1}$) below that of the acceptor and by holding the acceptor A concentration to the lowest possible value. The relative rates

$$\text{sens} + h\nu \longrightarrow {}^1(\text{sens})$$

$$^1(\text{sens}) \xrightarrow{k_{ic}} {}^3(\text{sens})$$

$$^1(\text{sens}) + A \xrightarrow{k_s} \text{sens} + {}^1A$$

$$^3(\text{sens}) + A \xrightarrow{k_t} \text{sens} + {}^3A$$

of intersystem crossing and singlet energy transfer will be given by the ratio

$$\frac{k_{ic}[{}^1(\text{sens})]}{k_s[{}^1(\text{sens})][A]} = \frac{k_{ic}}{k_s[A]}$$

The highest value k_s can have will be that of a diffusion controlled rate constant (10^9–10^{10}). If k_{ic} is as low as 10^8 sec^{-1}, it is easy for singlet energy transfer to become dominant. This is especially true if the concentration of the acceptor is high. In an acceptable sensitization experiment, the following conditions must be met in addition to the strictures given above. The sensitizer must absorb all the light or a known fraction of the light. The sensitizer triplet should transfer energy to the acceptor at a diffusion controlled rate. This condition will generally be met if the triplet energy of the sensitizer exceeds that of the acceptor by at least 3 kcal/mole. The sensitized reaction must give the same products in the same ratios as the unsensitized reaction. As an example, consider the photochemical rearrangement of 4,4-dimethyl-2-cyclohexenone in t-butyl alcohol, which gives two primary products in the same ratio in the direct irradiation and in the benzophenone sensitized process. Different ratios would imply some singlet reaction or involvement of an upper triplet excited state.

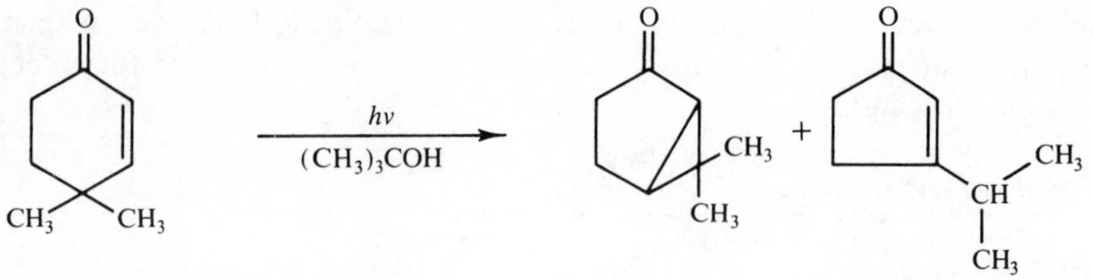

Direct irradiation		1	:	1.11
Benzophenone sensitized ($3660\overset{\bullet}{A}$)		1	:	1.08

5.6 QUENCHING

Quenching is another useful probe for determination of mechanism. It is possible to quench reactions and to quench light emission from excited states. Both kinds of quenching are important in the study of reaction mechanism. If we consider a simple mechanism of the type shown below, assumption of a steady-state concentration for the reactive

Process		Rate	
excitation	$M \xrightarrow{h\nu} M^*$	$I_0\phi$	I_0 = number of quanta absorbed by M per unit time
deactivation	$M^* \xrightarrow{k_d} M$	$k_d[M^*]$	
reaction	$M^* \xrightarrow{k_r}$ products	$k_r[M^*]$	ϕ = efficiency of production of the reactive excited state (M^*)
quenching	$M^* + Q \xrightarrow{k_q} M + Q^*$	$k_q[M^*][Q]$	

excited state (M^*) permits analysis of the kinetics of the system in terms of the quantum efficiency of the reaction in presence and absence of a quencher. The steady-state concentration of M^* will be achieved when the rate of production of M^* is equal to the sum of the rates of destruction. This relationship can be solved for the concentration of M^*. This value

$$I_0\phi = k_d[M^*] + k_r[M^*] + k_q[M^*][Q]$$

$$[M^*] = \frac{I_0\phi}{k_d + k_r + k_q[Q]}$$

can be used to obtain an expression for the quantum efficiency of the process. In the absence of a quencher, the term $k_q[Q]$ is zero. We can then

$$\Phi_{product} = \frac{\text{rate of product formation}}{\text{rate of absorption of quanta}}$$

$$\Phi_{product} = \frac{k_r[M^*]}{I_0} = \frac{k_r\phi}{k_d + k_r + k_q[Q]}$$

take a ratio of the quantum efficiency in the absence of quencher to that in the presence of a quencher. The resulting equation predicts a linear plot

$$\frac{(\Phi_{prod})_0}{(\Phi_{prod})_q} = \frac{k_r\phi/(k_d + k_r)}{k_r\phi/(k_d + k_r + k_q[Q])} = \frac{k_d + k_r + k_q[Q]}{k_d + k_r}$$

$$\frac{(\Phi_{prod})_0}{(\Phi_{prod})_q} = 1 + \frac{k_q[Q]}{k_d + k_r}$$

of the ratio of the quantum efficiencies for unquenched and quenched reactions (Φ_0/Φ_q) as function of the quencher concentration; this is known as a Stern-Volmer plot.

The rearrangement of 4,4-dimethyl-2-cyclohexenone may be used as an example of quenching of a reaction. Di-tertiarybutyl nitroxide is used as a quencher because it is an efficient quencher for triplet excited states

$$(CH_3)_3C\diagdown$$
$$N—O$$
$$(CH_3)_3C\diagup \quad \cdot$$

di-*t*-butyl nitroxide

and because it does not absorb strongly the 3130 Å light used in the irradiation of 4,4-dimethyl-2-cyclohexenone. This system gives a linear Stern-Volmer plot (Φ_0/Φ_q *vs.* [Q]). The slope of a Stern-Volmer plot is

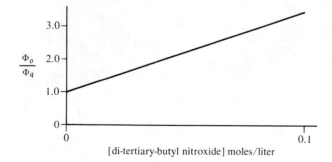

[di-tertiary-butyl nitroxide] moles/liter

equal to $k_q\tau$, where τ is the mean lifetime of the reactive excited state being quenched. The value of k_q can be estimated from the Debye equation for diffusion controlled rates. Quenching thus provides a kinetic method for estimating excited state lifetimes.

Quenching of light emission from an excited state can be useful in the study of mechanism, especially in identifying reactive singlet excited states. Irradiation of *trans*-stilbene in the presence of tetramethylethylene (4M) gives a 1:1 adduct with a quantum efficiency of 0.54 at room temperature. The photocycloaddition of *trans*-stilbene to tetramethylethylene competes with *trans* to *cis* isomerization of the stilbene. When the reaction is sensitized with thioxanthone, only *trans* to *cis* isomerization is observed. This implies, but does not prove, that it is the low-lying singlet

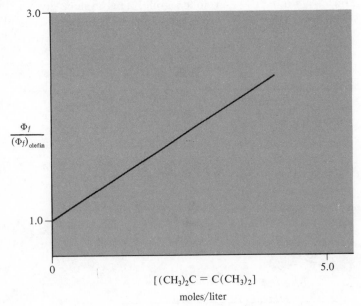

trans-stilbene   tetramethylethylene

excited state of *trans*-stilbene which adds to tetramethylethylene. Fortunately, *trans*-stilbene emits light from its low-lying singlet; i.e., it fluo-

$$\frac{\Phi_f}{(\Phi_f)_{\text{olefin}}}$$

$[(CH_3)_2C = C(CH_3)_2]$

moles/liter

Fig. 5-2

resces. If tetramethylethylene is adding to the S_1 state of *trans*-stilbene, it must quench the fluorescence. This is found to be true, and the quenching again follows the Stern-Volmer equation (Fig. 5-2). Notice that much higher quencher concentrations are required to quench the excited singlet state than were required for quenching the triplet excited state in 4,4-dimethyl-2-cyclohexenone. The quenching of *trans*-stilbene fluorescence by tetramethylethylene is a consequence of exciplex formation (see Chapter 7 for a detailed account of the mechanism of addition of *trans*-stilbene to olefins).

5.7 EMISSION SPECTROSCOPY

Fluorescence and phosphorescence provide important data on excited state properties such as lifetime and electronic configuration. It is often possible to correlate spectroscopic properties of molecules with their photochemical behavior. The phosphorescence from aromatic ketones has proved particularly useful in this regard. The phosphorescence from $^3(n,\pi^*)$ states is characterized by vibronic structure (see benzophenone phosphorescence) and short lifetimes. The phosphorescence from $^3(\pi,\pi^*)$ states is usually broad and featureless and has a longer lifetime (see *m*-methoxyacetophenone). If we consider the effect of substituents on the

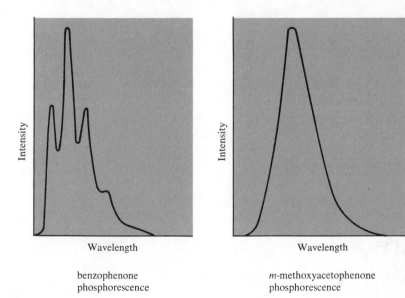

benzophenone
phosphorescence

m-methoxyacetophenone
phosphorescence

Fig. 5-3

$$\underset{\substack{\text{Ar}-\text{C} \\ \text{CH}_2}}{\overset{\text{O}}{\parallel}} \overset{\text{H}}{\underset{\text{CH}_2}{\overset{\text{CH}_2}{\diagdown}}} \xrightarrow{h\nu} \underset{\text{Ar}-\text{C}}{\overset{\text{O}}{\parallel}} \diagdown_{\text{CH}_3} + \quad \text{CH}_2 = \text{CH}_2$$

Ar	Φ_{II}	Life-time in msec (configuration)	
C_6H_5-	0.42	5	(n,π^*)
$p\text{-F}-C_6H_4-$	0.29	4	(n,π^*)
$m\text{-F}-C_6H_4-$	0.38	8	(n,π^*)
$m\text{-CH}_3O-C_6H_4-$	0.005	370	(π,π^*)

nature of the phosphorescence of aromatic ketones and their behavior in photoreduction or Norrish type II cleavage, the correlation is immediately apparent. Ketones with π, π^* configurations have long excited state lifetimes and are very unreactive in photo reductions (small quantum efficiency for reduction).

$$\underset{\text{Ar}-\text{C}-\text{CH}_3}{\overset{\text{O}}{\parallel}} \xrightarrow[\text{R H}]{h\nu} \underset{\substack{\text{CH}_3 \; \text{CH}_3}}{\overset{\text{OH OH}}{\underset{\big|\;\;\;\;\big|}{\text{Ar}-\text{C}-\text{C}-\text{Ar}}}}$$

Ar	$\Phi_{Red'n}$	life-time in msec. (configuration)	
C_6H_5	0.35	4.0	(n, π^*)
$p-CF_3-C_6H_4-$	0.72	0.7	(n, π^*)
$CH_3 \text{—⬡—} CH_3$	0.07	170	(π, π^*)
(methylenedioxyphenyl)	0.002	370	(π, π^*)
$CH_3O \text{—⬡—}$	0.006	250	(π, π^*)

5.8 PROBLEMS

1. Derive an expression for $1/\Phi_a$ as a function of $1/[O]$ for the mechanism given below. The quantum yield for photoaddition is Φ_a, and the olefin concentration is $[O]$.

$$K + h\nu \longrightarrow K^*$$

$$K^* \xrightarrow{k_d} K$$

$$K^* + O \xrightarrow{k_a} \text{product}$$

Hint:

$$\Phi = \frac{\text{rate of formation of product}}{\text{rate of absorption of quanta}}$$

What is the significance of the intercept in a plot of $1/\Phi_a$ *vs.* $1/[O]$? What is the significance of the slope?

2. In many examples of Norrish type II cleavage, both $^1(n,\pi^*)$ and $^3(n,\pi^*)$ states are reactive. Draw a schematic plot of Φ_0/Φ_q as a function of quencher concentration for the case of a quencher (such as di-*t*-butylnitroxide) that quenches both singlet and triplet excited states.

REFERENCE

J. G. Calvert and J. N. Pitts, Jr., *Photochemistry*. New York: Wiley, 1966, Chapters 6 and 7.

6

Molecular Orbital Symmetry and the Stereochemistry of Concerted Unimolecular Reactions

6.1 INTRODUCTION

In earlier chapters we have emphasized the differences between thermal and photochemical processes; we now turn our attention to features that they have in common. Many concerted reactions, thermal and photochemical, show a high degree of stereospecificity. Furthermore, it is often observed that related thermal and photochemical processes, while both stereospecific, give products with differing stereochemistry. In this chapter we shall explore a molecular orbital view of the stereochemistry of electrocyclic reactions and sigmatropic reactions.

The chemistry related to vitamin D shows many stereospecific concerted reactions, both thermal and photochemical. It is particularly striking that previtamin D undergoes a photochemical cyclization to ergosterol, which has the hydrogen atom at position 9 and the methyl group at position 10 *trans*, while the thermal cyclization gives pyrocalciferol and isopyrocalciferol, which have the substituents at positions 9 and 10 *cis*. The photochemical reactions of the diene systems in pyrocalciferol and isopyrocalciferol are not like the conversion of ergosterol to previtamin D. Instead, isomerization to cyclobutene derivatives is observed.

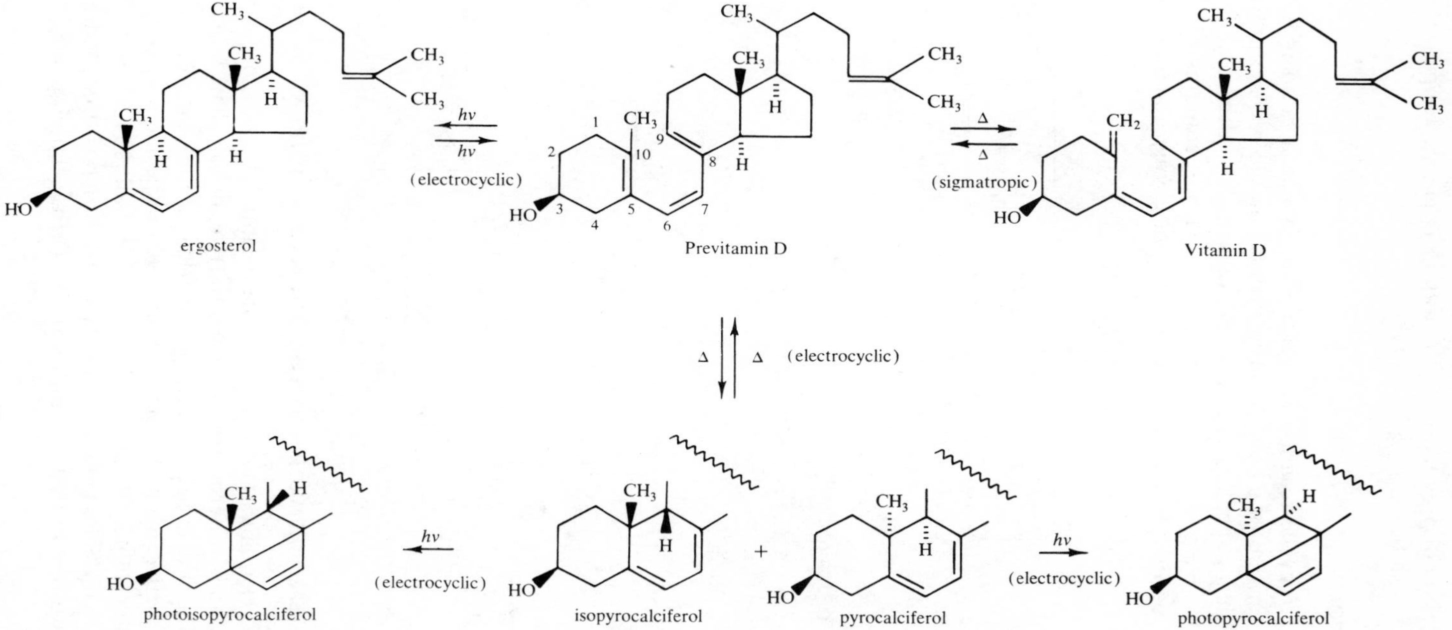

Fig. 6-1

The conversion of previtamin D to vitamin D is an example of a thermal 1,7-hydrogen shift.

6.2 ELECTROCYCLIC REACTIONS

An electrocyclic reaction is a process in which a bond is formed between the termini of a conjugated π system, or the reverse process. It is thus possible to speak of electrocyclic closure and electrocyclic opening of rings. If the π system of the open-chain partner contains $k\pi$ electrons, the cyclic partner will contain $(k - 2)\pi$ electrons and one additional σ bond.

6.3 THE STEREOCHEMISTRY OF ELECTROCYCLIC REACTIONS

Let us examine the closure of previtamin D to ergosterol (photochemical) and isopyrocalciferol (thermal) in detail. The important elements of the system are the conjugated triene in the starting material and the diene in the products. Since similar stereospecific reactions occur in other cyclohexadiene-hexatriene systems, we shall reduce the system to the model shown. If we view the molecules from a position in the plane of the ring, we realize that a rotation of the carbon holding the substituent

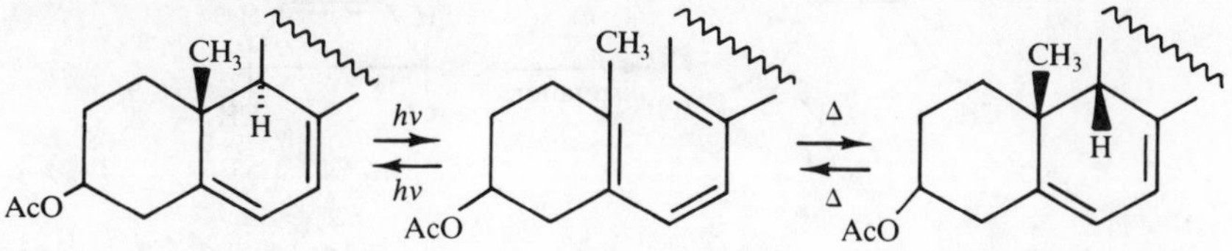

model system

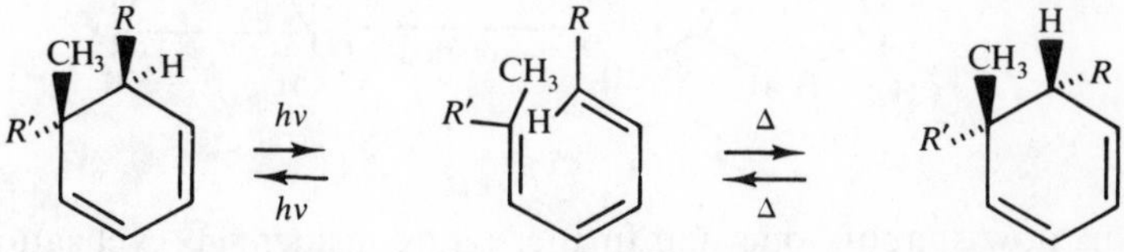

groups must take place, since these substituents lie in a plane approximately perpendicular to the plane of the ring carbons in the starting diene, and in the same plane as the carbons of the triene in the product. Both terminal carbons may rotate in the same direction (conrotation) or in opposite directions (disrotation). Conrotation is observed in the photochemical opening and closing; disrotation is observed in the thermal

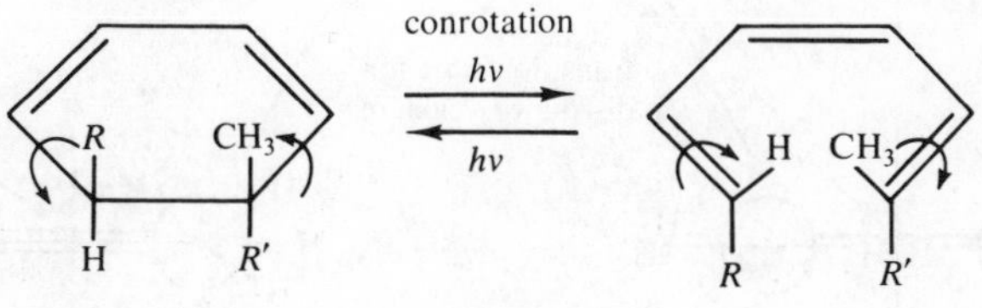

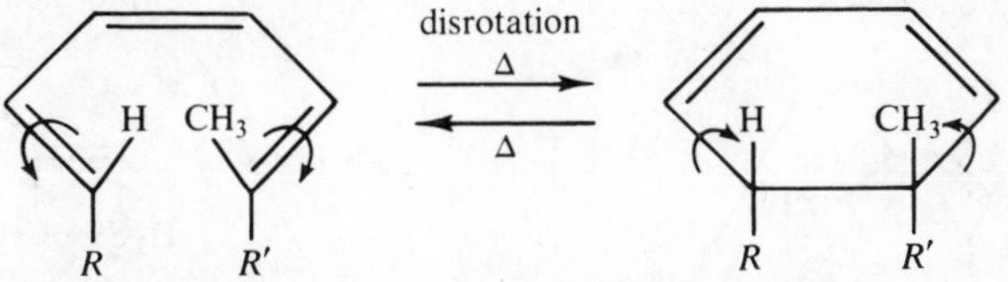

conrotation
hv

disrotation
Δ

opening and closing. The two electrocyclic processes, thermal and photochemical, thus follow opposite stereochemical courses. Two modes of conrotation and two modes of disrotation are possible in the model system, and application of the opposite mode would in each case lead to different products. In the steroid systems the constraint of the additional rings is responsible for the single mode of conrotation and disrotation observed.

Consider now the process in which a 1,3-butadiene system cyclizes to a cyclobutene (isopyrocalciferol to photoisopyrocalciferol, for example). In this case the photochemical cyclization is disrotatory, and the thermal cyclization is conrotatory.

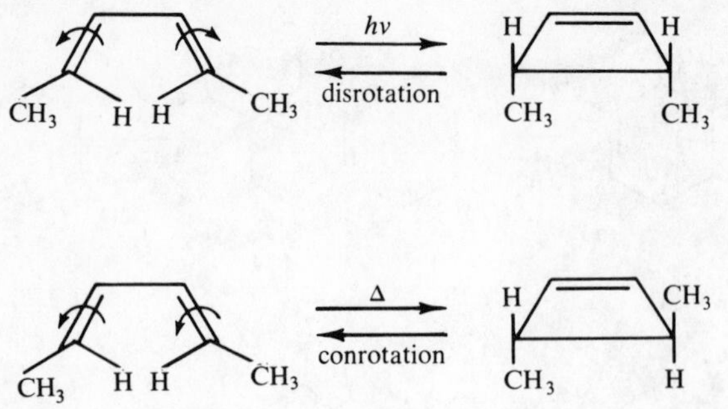

Let us now simplify one step further and consider the cyclization of 1,3-butadiene. The symmetry properties of cisoid butadiene are such that it

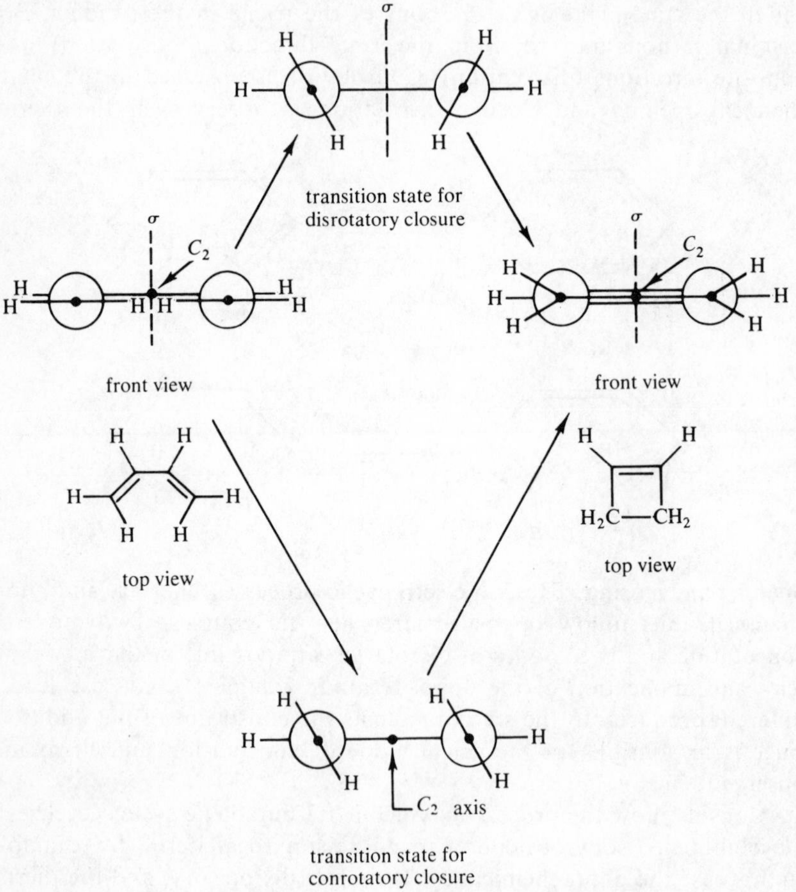

transition state for
disrotatory closure

σ C_2

front view

front view

top view

top view

transition state for
conrotatory closure

has a plane of symmetry (σ) perpendicular to the plane of the molecule and bisecting the carbon-carbon single bond and a C_2 axis of symmetry that lies in the plane of the molecule and is equidistant from carbons 1 and 4 and carbons 2 and 3. A C_n axis is known as an n-fold axis of symmetry. A C_n axis is defined as an axis such that rotation about this axis through $360°/n$ gives an identity. In the disrotatory photochemical cyclization the molecule maintains the symmetry plane throughout the course of the reaction, while the thermal cyclization maintains a C_2 axis. These symmetry properties are characteristic of conrotation and disrotation in unsubstituted conjugated polyene cyclizations. A C_2 axis of symmetry is maintained in the conrotatory photochemical cyclization of 1,3,5-hexatriene, and a plane of symmetry is maintained in the disrotatory thermal cyclization.

6.4 SYMMETRY PROPERTIES OF MOLECULAR ORBITALS

The wave function for the π orbital of ethylene has no nodes, whereas the wave function for the π^* orbital has one node. We need a convenient representation for the wave functions that shows the relative signs of the wave function at different parts of the molecule. For this purpose we shall use an atomic orbital representation with relative signs shown by light and dark orbital lobes corresponding to the sign of the wave function.

The π orbital of ethylene has a plane of symmetry σ, which bisects the molecule and is perpendicular to the molecular plane. The π^* orbital has a

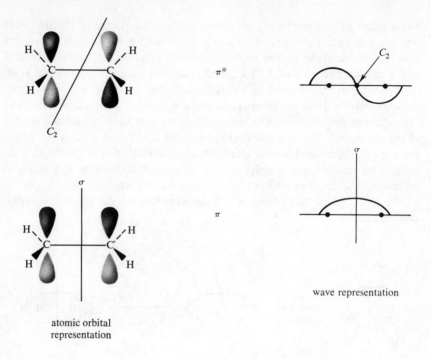

π^*

σ

π

wave representation

atomic orbital
representation

2-fold symmetry axis C_2 in the plane of the molecule, perpendicular to the carbon-carbon bond and bisecting the molecule. The π orbital does not have the C_2 axis of symmetry defined above for π^*, and the π^* orbital does not have the plane of symmetry specified above for the π orbital. The two orbitals thus transform differently on application of the symmetry operations σ (reflection in a plane) and C_2 (180° rotation about a specified axis). An orbital is symmetric with respect to a symmetry operation if application of the symmetry operation produces an identity and antisymmetric if it does not. The π orbital of ethylene is symmetric S with respect to reflection in the plane σ and antisymmetric A with respect to rotation about the C_2 axis. The π^* orbital of ethylene is antisymmetric with respect to σ and symmetric with respect to C_2.

	SYMMETRY ELEMENT	
	σ	C_2
π^*	Antisymmetric (A)	Symmetric (S)
π	Symmetric (S)	Antisymmetric (A)

The π orbitals of butadiene can also be characterized as symmetric or antisymmetric with respect to the plane of symmetry and the C_2 axis of symmetry shown. The results are shown in Table 6.1.

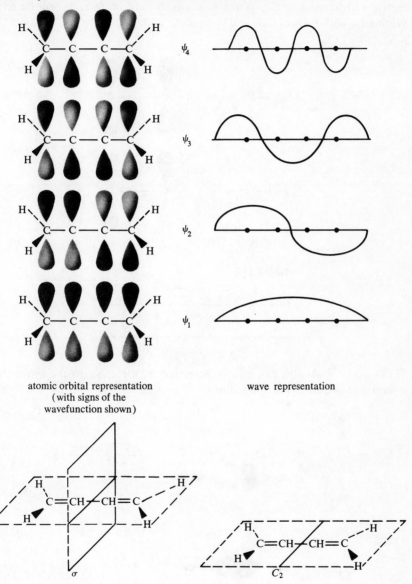

atomic orbital representation (with signs of the wavefunction shown)

wave representation

Table 6.1

	σ	C_2
ψ_4	A	S
ψ_3	S	A
ψ_2	A	S
ψ_1	S	A

These results can now be generalized. For any linear conjugated π system, the wave function ψ_n will have $n - 1$ nodes. When $n - 1$ is zero or an even integer, ψ_n will be symmetric with respect to σ and antisymmetric with respect to C_2. When $n - 1$ is an odd integer, ψ_n will be antisymmetric with respect to σ and symmetric with respect to C_2.

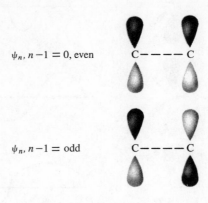

$\psi_n, n - 1 = 0$, even

$\psi_n, n - 1 =$ odd

Table 6.2

	σ	C_2
$\psi_n, n - 1 = 0$, even	S	A
$\psi_n, n - 1 =$ odd	A	S

Sigma orbitals also can be characterized with respect to the symmetry operations (σ and C_2) defined above.

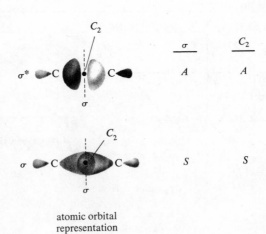

	σ	C_2
σ^*	A	A
σ	S	S

atomic orbital representation

6.5 SYMMETRY CONTROL OF ELECTROCYCLIC REACTIONS

We are now in a position to consider an electrocyclic reaction in terms of orbital symmetry. Consider the disrotatory conversion of cyclobutene to butadiene. The orbitals that undergo direct changes in cyclobutene are σ, π, and the related antibonding orbitals σ^* and π^*, and in butadiene ψ_1, ψ_2, ψ_3, and ψ_4. A plane of symmetry σ is maintained throughout the course of the reaction. In a concerted reaction, it is required that orbital symmetry be conserved throughout. This means that a symmetric orbital in the starting material must transform into a symmetric orbital in the product and that an antisymmetric orbital must transform into an antisymmetric orbital.

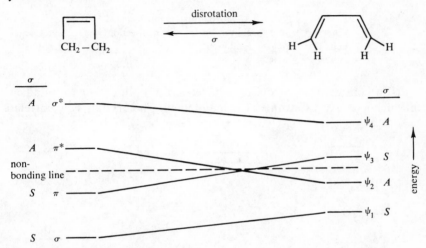

The orbitals that correlate (transform into each other) are connected by lines (see above). Useful information is immediately available from this correlation diagram. First, the cyclobutene ground state $\sigma^2\pi^2$ correlates with a doubly excited state $\psi_1^2\psi_3^2$ of butadiene.[†] Doubly excited states (two electrons promoted to higher-energy orbitals) are very high-energy states, and this is clearly an unfavorable process. In orbital symmetry terms, this process is forbidden. The ground state of the starting material must correlate with the ground state of the product for a thermal reaction to be symmetry-allowed. The first excited state of starting material must correlate with the first excited state of the product for a photochemical reaction to be allowed. Further inspection shows that the first excited state of cyclobutene $\sigma^2\pi\pi^*$ correlates with the first excited state of butadiene

[†] Only the orbitals that undergo change are cited in this description of the ground and excited state configurations. These descriptions are analogous to those used for atoms, e.g., the ground state carbon atom, $1s^2 2s^2 2p^2$.

$\psi_1^2\psi_2\psi_3$. The disrotatory process is photochemically allowed (in either direction).

Correlation		Conclusion
$\sigma^2\pi^2$ ground state	$\rightarrow$ $\psi_1^2\psi_3^2$ upper excited state	Disrotatory thermal conversion of cyclobutene to butadiene is forbidden.
$\psi_1^2\psi_2^2$ ground state	$\rightarrow$ $\sigma^2\pi^{*2}$ upper excited state	Disrotatory thermal conversion of butadiene to cyclobutene is forbidden.
$\sigma^2\pi\pi^*$ first excited state	$\rightarrow$ $\psi_1^2\psi_2\psi_3$ first excited state	A disrotatory photochemical conversion in either direction is allowed.

Now consider the conrotatory conversion of cyclobutene to butadiene in which a C_2 axis of symmetry is maintained. The orbitals now correlate

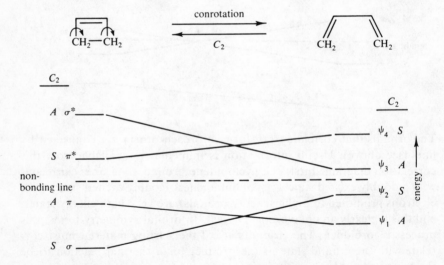

in such a way that the ground state of cyclobutene $\sigma^2\pi^2$ correlates with the ground state of butadiene $\psi_1^2\psi_2^2$. The thermal conrotatory process is thus allowed in either direction. The first excited state of cyclobutene correlates with an upper excited state of butadiene. The photochemical conrotatory opening is thus forbidden. A similar argument shows that the photochemical conrotatory closure of butadiene is also forbidden.

Correlation	Conclusion
$\sigma^2\pi^2 \rightarrow \psi_1^2\psi_2^2$ ground state	ground state	Thermal conrotatory process in either direction is allowed.
$\sigma^2\pi\pi^* \rightarrow \psi_1\psi_2^2\psi_4$ first excited state	upper excited state	Photochemical conrotatory opening of cyclobutene to butadiene is forbidden.
$\psi_1^2\psi_2\psi_3 \rightarrow \sigma\pi^2\sigma^*$ first excited state	upper excited state	Photochemical conrotatory closure of butadiene to cyclobutene is forbidden.

The correlation diagrams shown in the preceding diagram and the diagram on page 93 provide two very practical conclusions. First, the thermal opening of a cyclobutene should be a conrotatory process. Second, the photochemical closure of butadiene to cyclobutene should be a disrotatory process. These predictions are in accord with experiment. Thermal isomerization of *cis*-3,4-dimethylcyclobutene gives *cis,trans*-2,4-

cis-3, 4-dimethyl-
cyclobutene

Δ
conrotation

cis, trans-2,4-hexadiene

cis-bicyclo-
[6.2.0] deca-2, 9-diene

Δ
conrotation

trans, cis, cis-cyclo-
deca-1, 3,5-triene

cis, cis, cis-2, 4, 6-
cyclooctatrienone

hv

trans, cis, cis-2, 4, 6-
cyclooctatrienone

Δ

cis-bicyclo [4.2.0]-
octa-3, 6-dien-2-one

hexadiene, and *cis*-bicyclo[6.2.0]deca-2,9-diene gives *trans,cis,cis*-1,3,5-cyclodecatriene. Photochemical isomerization of *cis,cis,cis*-2,4,6-cyclooctatrienone gives *trans,cis,cis*-2,4,6-cyclooctatrienone, which cyclizes thermally in a conrotatory process to *cis*-bicyclo[4.2.0]octa-3,6-dien-2-one.

The photochemical disrotatory ring closure reactions of isopyrocalciferol and pyrocalciferol (page 85) are also in agreement with prediction.

None of the examples given above maintains the requisite symmetry for construction of a correlation diagram. In general, we must simplify our structures by ignoring substituents to permit detailed analysis of the system. Our assumption is thus that the analysis of the simple parent system will apply to substituted derivatives as well.

Examples are known that do not fit the orbital symmetry arguments developed above. Bicyclo[3.2.0]hept-6-ene, for example, isomerizes thermally to *cis,cis*-cyclohepta-1,3-diene. This reaction requires rather

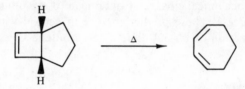

cis-bicyclo [3.2.0] hept-6-ene   cis, cis-cyclohepta-1, 3-diene

high temperature and is probably not concerted. Orbital symmetry arguments tell us what can happen by a concerted path, but not all reactions occur via concerted paths. Any thermally forbidden reaction may occur under sufficiently vigorous conditions. One is thus led to ask how much difference in energy exists between an orbital symmetry-allowed process and an orbital symmetry-forbidden process in the same system. Estimates of the difference in activation energies for allowed and forbidden paths fall in the range 10–15 kcal/mole. This impressive difference is certainly large enough to determine the course of most reactions.

The correlation diagrams show why orbital symmetry controls the stereochemical course of concerted reactions, but they require some little time to construct. We now turn to a simple method for making rapid predictions. This approach is based on the postulate that *the stereochemistry of an electrocyclic process is determined by the symmetry of the highest occupied molecular orbital (HOMO) of the open-chain partner.* If the highest occupied molecular orbital has the plane of symmetry σ, the process will be disrotatory. If the highest occupied molecular orbital has the axis (C_2) of symmetry, the process will be conrotatory. The reasons behind this rule can be understood by recalling that overlap of wave functions of the same sign is bonding, whereas overlap of wave functions of opposite sign is antibonding.

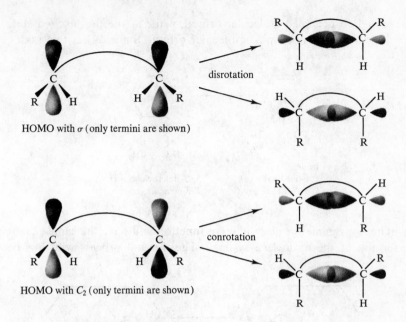

HOMO with σ (only termini are shown)

disrotation

HOMO with C_2 (only termini are shown)

conrotation

Now we are in a position to consider specific examples of the application of this rule. Consider the stereochemistry of the thermal and photochemical closure of *trans,cis,trans*-1,6-dimethylhexa-1,3,5-triene to 5,6-dimethylcyclohexa-1,3-diene. Remember that the rule states that the orbital of interest (HOMO) is in the *open-chain partner*, i.e., the hexatriene. We have six π electrons in hexatriene; in the ground state ψ_3 will

ψ_6 ——   —— ψ_6

ψ_5 ——   —— ψ_5

ψ_4 —— $\xrightarrow{h\nu}$ $\underset{}{+}$ ψ_4 HOMO

ψ_3 $\underset{}{+\!\!\!+}$ HOMO $\underset{}{+}$ ψ_3

ψ_2 $+\!\!\!+$ $+\!\!\!+$ ψ_2

ψ_1 $+\!\!\!+$ $+\!\!\!+$ ψ_1

ground   first excited
state   state

$\psi_1{}^2\psi_2{}^2\psi_3{}^2$ $\psi_1{}^2\psi_2{}^2\psi_3\psi_4$

hexatriene

be the highest occupied molecular orbital, while in the first excited state ψ_4 will be the highest occupied molecular orbital. Since ψ_3 has two nodes,

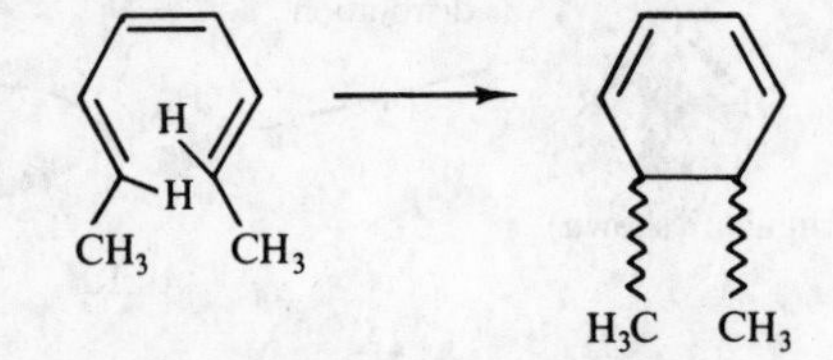

trans, cis, trans-
1, 6-dimethylhexa-
1, 3, 5-triene

5, 6-dimethylcyclohexa-
1, 3-diene

it will have σ symmetry; i.e., the wave function will have the same sign at the termini of the hexatriene system. The thermal process will thus be

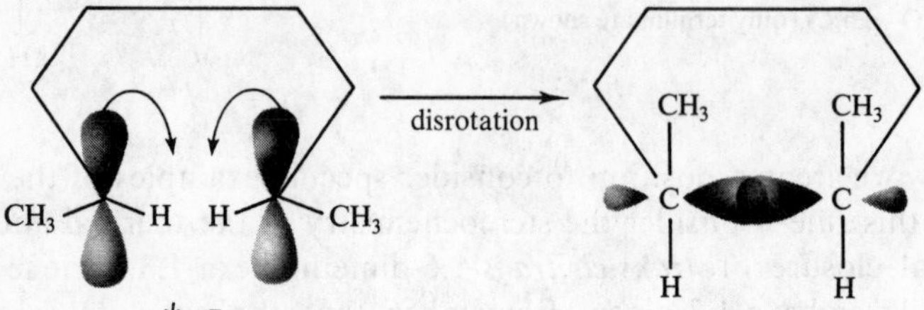

ψ_3, σ
(only signs at termini are shown)

disrotatory and will lead to *cis*-5,6-dimethylcyclohexa-1,3-diene. The stereochemistry of the photochemical process will be governed by ψ_4, which has three nodes and consequently C_2 symmetry. The photochem-

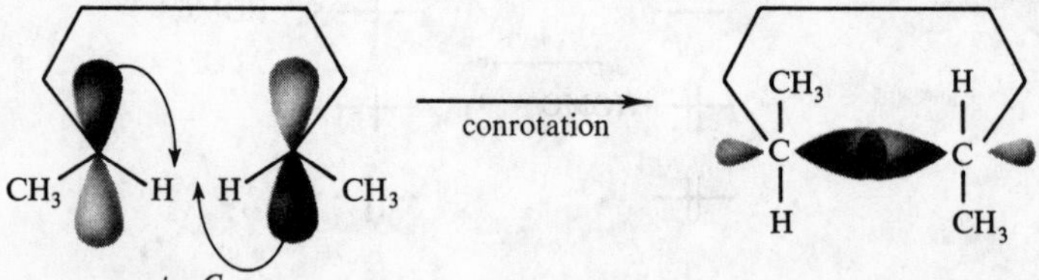

ψ_4, C_2
(only signs at termini are shown)

ical process thus will be conrotatory and will lead to *trans*-5,6-dimethyl-cyclohexa-1,3-diene.

Now let us consider the electrocylic opening of *cis*-7,8-dimethylcyclo-octa-1,3,5-triene. The ground state configuration of the open-chain octatetraene will be $\psi_1^2\psi_2^2\psi_3^2\psi_4^2$ in which ψ_4 is the HOMO, and the configuration of the first excited state will be $\psi_1^2\psi_2^2\psi_3^2\psi_4\psi_5$, in which ψ_5 is the HOMO. The cyclooctatriene ring must open in such a fashion that the σ bond orbitals transform into the highest occupied molecular orbital in each case.

ψ_5, 4 nodes, σ
(only signs at termini are shown)

favored for
steric reasons

The thermal process is conrotatory with "either possible conrotation" giving the same product. The photochemical process is disrotatory with two products possible. The disrotation shown with dotted arrows leads to severe steric interactions between the methyl groups. This interaction is avoided in the alternate process in which the methyl groups move away from each other, and this is the favored process.

Table 6.3 summarizes a modest number of orbital symmetry predictions. Many of these have been confirmed by experiment.

6.6 SIGMATROPIC REACTIONS

Now that we have a basis for discussing orbital symmetry control of stereochemistry in concerted reactions, we turn our attention to sigmatropic reactions. We have already encountered two examples of this type

Table 6.3

STEREOCHEMISTRY OF ELECTROCYCLIC REACTIONS

Open chain isomer	Conditions	Symmetry of HOMO (ψn)	Mode of closure or opening	Cyclic partner
	Δ	$\sigma(\psi_1)$	disrotation	
	hv	$C_2(\psi_2)$	conrotation	
	Δ	$C_2(\psi_2)$	conrotation	
	hv	$\sigma(\psi_3)$	disrotation	
	Δ	$C_2(\psi_2)$	conrotation	
	hv	$\sigma(\psi_3)$	disrotation	
	Δ	$C_2(\psi_2)$	conrotation	
	hv	$\sigma(\psi_3)$	disrotation	
	Δ	$\sigma(\psi_3)$	disrotation	
	hv	$C_2(\psi_4)$	conrotation	
	Δ	$C_2(\psi_4)$	conrotation	
	hv	$\sigma(\psi_5)$	disrotation	

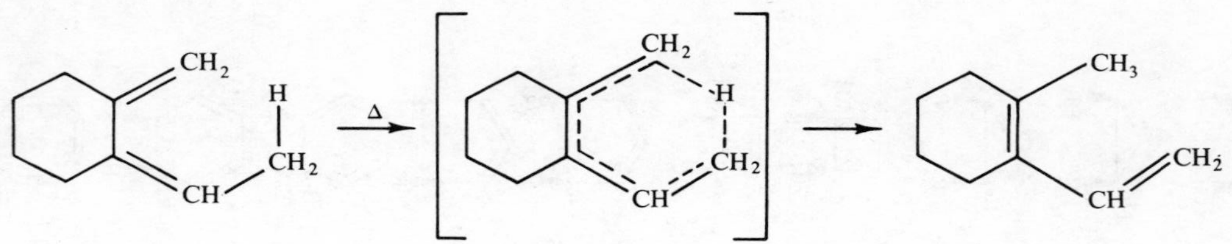

transition state

sigmatropic reaction of order [1, 5]

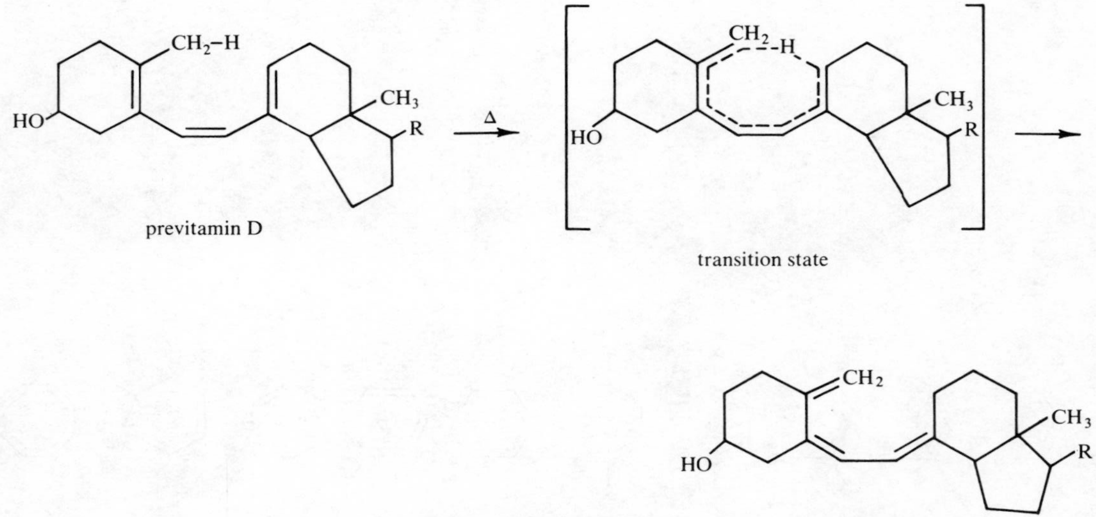

transition state

previtamin D

sigmatropic reaction of order [1, 7]

vitamin D

of reaction, involving concerted transfer of a hydrogen atom from one end of a conjugated system to another. The first example is a sigmatropic reaction of order [1,5], the second of order [1,7]. In general, a sigmatropic reaction of order $[i,j]$ is defined as the migration of a σ bond, flanked by one or more π-electron systems, to a new position whose termini are $i-1$ and $j-1$ atoms removed from its original bonded loci, in an uncatalyzed intramolecular process. In the first case cited above, the C—H σ bond moves from the 1,1 position to the 1,5 position, in the second from 1,1 to 1,7. The numbers refer to the *atoms* at either end of the sigma bond which is thought of as moving. In the Cope rearrangement, both ends of the bond are attached in new positions in the product. The Cope rearrangement (Chapter 2) is an example of a sigmatropic rearrangement of order [3,3]. We can imagine sigmatropic shifts of various other orders, [1,3], [3,5], etc.

transition state

sigmatropic reaction of order [3, 3]

There are two stereochemically different courses by which sigmatropic processes may occur. In one stereochemical course the hydrogen atom (or other migrating group) remains on the same side of the molecular plane throughout the course of the reaction. This is known as a *suprafacial* shift, and in the transition state there is a plane of symmetry σ. In the

transition state
(σ symmetry)

suprafacial sigmatropic reaction of order [1, 5]

other stereochemical course, the group R is transferred from the top face of the carbon to which it was originally attached to the bottom face of the carbon to which it is migrating;

antarafacial sigmatropic reaction of order [1, 5]

This mode of transfer is called *antarfacial* transfer, and the transition state is characterized by C_2 symmetry.

We cannot draw correlation diagrams for a sigmatropic reaction as we did for an electrocyclic reaction, because symmetry is not maintained throughout the reaction, although it does appear in the transition state. The starting material, for example, does not have the σ or C_2 symmetry present in the transition state. We can, however, correctly predict the stereochemical course of sigmatropic reactions on the basis of the symmetry properties of the highest occupied molecular orbital of the reacting system. In order to do this, we shall have to use as our model for the system a structure that is related to the transition state. Let us imagine, then, that the C—H bond that is migrating is made up of a hydrogen $1s$ orbital and a carbon $2p$ orbital (this is a useful fiction). One electron is assigned to the hydrogen $1s$ orbital and one to the carbon $2p$ orbital. Thus for a [1,5] hydrogen shift we have the following possibilities, ignoring for the moment orbital signs.

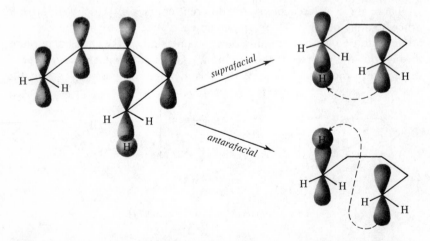

In our model we have, then, a π system made up of five $2p$ atomic orbitals and five electrons (the pentadienyl radical). The configuration of the ground state for such a system will be $\psi_1^2\psi_2^2\psi_3$, and the HOMO will be ψ_3. The first excited state will have the configuration $\psi_1^2\psi_2^2\psi_4$

(ψ_3 is vacant), and the HOMO will be ψ_4. The thermal process is controlled by the σ symmetry of ψ_3, which corresponds to the symmetry of

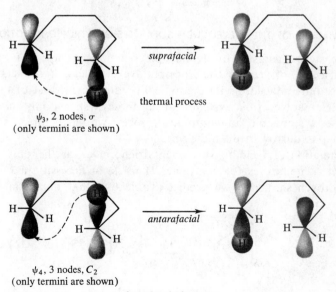

pentadienyl radical

the *suprafacial* transition state. The photochemical process is governed by the C_2 symmetry of ψ_4 which corresponds to the symmetry of the *antarafacial* transition state. The relationship between the symmetry of ψ_3 and ψ_4 and the *suprafacial* or *antarafacial* course of the reaction is easily visualized as shown below.

An analysis of the Cope rearrangement (order [3,3]) and other similar reactions is slightly more complicated, because we have to consider the stereochemistry of both ends of the migrating bond. In a hydrogen migration we can, of course, ignore the stereochemistry around hydrogen. The Cope rearrangement (Secs. 2.2–2.5) proceeding through either a boat or chair transition state follows a *suprafacial* course with respect to each three-carbon fragment. Such a transformation is said to be *suprafacial-suprafacial*. The selection rules for any sigmatropic reaction of order [i,j] can be expressed in the following manner. For cases in which $i + j = 4Q$ (Q is an integer 1,2,...) the thermal reaction will be *antarafacial-suprafacial* or *suprafacial-antarafacial*, and the photochemical reaction will be *suprafacial-suprafacial* or *antarafacial-antarafacial*. For those cases in which $i + j = 4Q + 2$, the rules are reversed, and the thermal reactions are *suprafacial-suprafacial* or *antarafacial-antarafacial* and the photochemical reaction will be *antarafacial-suprafacial* or *suprafacial-antarafacial*.

Table 6.4

PREDICTED STEREOCHEMICAL COURSE FOR SIGMATROPIC REACTIONS

i,j	Thermal Reaction	Photochemical Reaction
1,3	*antarafacial*	*suprafacial*
1,5	*suprafacial*	*antarafacial*
1,7	*antarafacial*	*suprafacial*
3,3	*suprafacial-suprafacial* or *antarafacial-antarafacial*	*suprafacial-antarafacial*

6.7 EXAMPLES OF THE STEREOCHEMISTRY OF SIGMATROPIC REACTIONS

The stereochemistry of sigmatropic reactions is not as fully documented as that of electrocyclic processes. Nevertheless, those facts that we have are fully in accord with theoretical predictions. Most of the better examples involve cyclic systems for in these systems only *suprafacial* processes are possible; an *antarafacial* process would require the migrating group to burrow through the ring.

Irradiation of 7-methoxycycloheptatriene gives, as the only primary product, 1-methoxycycloheptatriene. This is, of necessity, a *suprafacial* 1,7 hydrogen shift, and we predict (Table 6.4) that this is an allowed

suprafacial

photochemical process. What happens when 7-methoxycycloheptatriene is heated? [1,7] or [1,3] thermal shifts are predicted to be *antarafacial*, and so should not occur because of the ring. A [1,5] *suprafacial* shift is allowed thermally, however, and this is the process observed. The boat-shaped conformation of cycloheptatriene is particularly suitable for such hydrogen transfers.

Perhaps the most dramatic example of orbital symmetry control of uncatalyzed unimolecular reactions is the thermal isomerization of 1-methoxybicyclo[3.2.0]hepta-3,6-dien-2-one to 3-methoxybicyclo[3.2.0]-hepta-3,6-dien-2-one. This is a most remarkable transformation, when one considers that the starting bicyclic ketone is a strained isomer of the tropolone system. The strained bicyclic isomer could relieve about 10 kcal/mole strain energy merely by opening to the tropolone. In addition, the tropolone system is a highly delocalized system (resonance energy

1-methoxybicyclo-
[3.2.0] hepta-3, 6-dien-2-one

3-methoxybicyclo-
[3.2.0] hepta-3, 6-dien-2-one

α-tropolone methyl ether

about 25–35 kcal/mole). There is thus at least 35 kcal/mole driving force for opening the bicyclic ketone to α-tropolone methyl ether. The opening, however, is a disrotatory opening of a cyclobutene to a butadiene, which is thermally forbidden. The conrotatory opening would lead to a *trans*-double bond in the α-tropolone methyl ether, and this is impossible if all trigonally hybridized atoms are to be even approximately coplanar. The conservation of orbital symmetry thus forbids what appears by any other argument to be an exceptionally favorable reaction. On the positive side, molecular orbital symmetry arguments suggest that the bicyclic ketone, which is a 1,5-hexadiene derivative, could undergo a thermally allowed *antarafacial-antarafacial* [3,3] sigmatropic reaction (a Cope rearrangement). This is, apparently, what happens. The 1,5-hexadiene system is made up of atoms in the following sequence: 6,7,1,5,4,3. The numbers show the shift of atoms. The movement of atoms has been demonstrated by deuterium labeling experiments. The *antarafacial-antarafacial* nature of

the rearrangement is more difficult to visualize. Suppose that we depict the changes with atomic orbitals using shaded lobes for reference. The bond formed between carbons 3 and 6 clearly is formed under the

molecule while the disrotation which breaks the 1,5-bond leads to 1,7- and 4,5-bond formation on the top side of the molecule. The bonding is thus *antarafacial* on the three-carbon fragment (6,7,1) and on the other three-carbon fragment (5,4,3) as well. This example illustrates clearly the utility of orbital symmetry arguments in understanding and predicting the course of concerted organic chemical reactions.*

The thermal isomerization of *endo*-6-acetoxybicyclo [3.2.0] hept-2-ene

---

* This assumes the reaction is concerted. A stepwise reaction has not been ruled out.

endo-6-acetoxybicyclo-
[3.2.0] hept-2-ene

exo-5-acetoxybicyclo-
[2.2.1] hept-2-ene

to *exo*-5-acetoxybicyclo [2.2.1] hept-2-ene appears, at first, to be a forbidden [1,3] *suprafacial* sigmatropic rearrangement. The rearrangement, however, proceeds with inversion at the carbon atom that moves. This inversion is shown by a deuterium labeling experiment. The transforma-

inversion at C_7

tion with inversion is allowed because inversion permits smooth overlap of the orbital of the carbon undergoing inversion with the HOMO (ψ_2) of the three-carbon unit, an allyl radical.

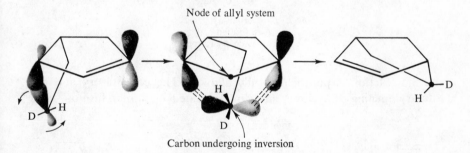

Node of allyl system

Carbon undergoing inversion

6.8 AN ALTERNATE QUALITATIVE MOLECULAR ORBITAL APPROACH

Electrocyclic reactions and sigmatropic reactions can be treated in the same fashion if we use a different approach to qualitative molecular orbital arguments. Consider a cyclic array of atomic orbitals that represent the orbitals undergoing change in the transition state and assign signs to the wave functions in the best manner for overlap. Count the number of nodes in the array and the number of electrons involved. An array with zero (or an even number) of nodes is stable in the ground state with $4n + 2$ electrons and in the excited state with $4n$ electrons. An array with an odd number of nodes is stable in the ground state with $4n$ electrons and in the excited state with $4n + 2$ electrons.† Consider the disrotatory cyclization of *cis*-1,3,5-hexatriene. The transition state can be

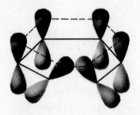

disrotatory cyclization
0 nodes, 6 electrons
Δ allowed

described by a cyclic array of atomic orbitals with no nodes and six electrons. This is a stable ground state system, and the reaction is thermally allowed. The conrotatory cyclization has an array with one node and six electrons. This arrangement is stable as an excited state,

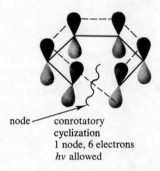

node —— conrotatory
cyclization
1 node, 6 electrons
hv allowed

and the reaction is photochemically allowed. The conrotatory and disrotatary opening of cyclobutenes can be treated in similar fashion. The

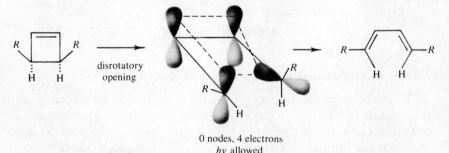

0 nodes, 4 electrons
hv allowed

†The ground state stability of arrays with zero nodes and $(4n + 2)$ π electrons is related to Hückel aromatic systems. Arrays with one node correspond to Möbius π systems that have a single twist in the π system and are stable (in theory) in the ground state with $4n$ π electrons.

transition state for disrotatory opening of a cyclobutene has zero nodes and four electrons in the cyclic array of orbitals. This arrangement is stable in the excited state, and the reaction is photochemically allowed. The conrotatory opening of cyclobutenes leads to a transition state with one node and four electrons, and is a thermally allowed process.

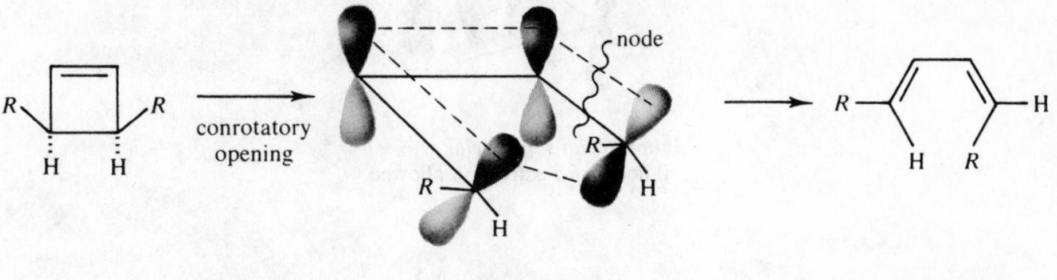

1 node, 4 electrons
Δ allowed

The same formalism can be used to treat sigmatropic reactions. The [1,3] *suprafacial* shift occurs via a transition with zero nodes and four

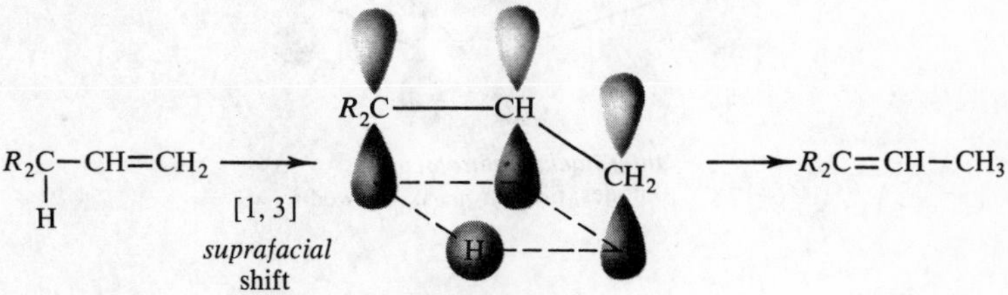

0 nodes,
4 electrons
hv allowed

electrons and thus is a photochemically allowed process. The Cope rear-

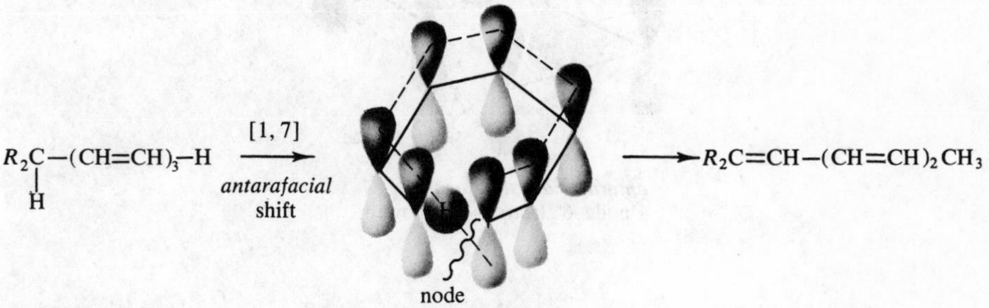

node

1 node, 8 electrons
Δ allowed

rangement in any of its stereochemical variations can also be treated by this formalism. Only transition state arrays are shown. The *suprafacial-*

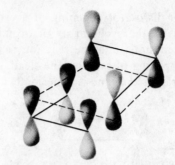

suprafacial-suprafacial
0 nodes, 6 electrons Δ allowed

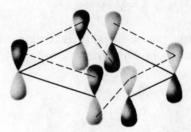

antarafacial-antarafacial
0 nodes, 6 electrons Δ allowed

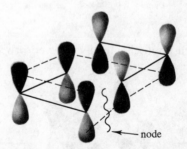

← node

antarafacial-suprafacial
1 node, 6 electrons *hv* allowed

suprafacial and *antarafacial-antarafacial* processes have arrays with zero nodes, whereas the *antarafacial-suprafacial* process has one node. Six electrons are involved in each case. The zero node transition states thus are favored in thermal reactions and the one-node transition state is favored in photochemical reactions.

6.9 PROBLEMS

1. Predict the stereochemistry of the following electrocyclic reactions.

2. Write a mechanism using an electrocyclic process for the reaction shown below.

3. Draw an orbital correlation diagram for the electrocyclic reaction shown below. Check your predictions against those in Table 6.3.

REFERENCES

1. R. B. Woodward and R. Hoffmann, *The Conservation of Orbital Symmetry*, New York: Academic Press, 1970.

2. R. B. Woodward, "Aromaticity," Special Publication No. 21, The Chemical Society, London, 1967, p. 217.

3. R. B. Woodward and R. Hoffmann, "Conservation of Orbital Symmetry," *Accounts of Chemical Research*, **1**, (1968), 17.

4. H. E. Zimmerman, *Angewante Chemie International Edition* (English), **8**, (1969), 1.

Cycloaddition

7.1 INTRODUCTION

Cycloaddition processes are among the most useful organic chemical reactions. The Diels-Alder reaction, known to every serious student of organic chemistry, is only one example of a great class of thermal and photochemical reactions. The facile synthesis of cyclic compounds from acyclic precursors, the high degree of stereo-selectivity, and subtle questions of mechanism all contribute to our interest in cycloaddition processes.

7.2 CLASSIFICATION OF CYCLOADDITION PROCESSES

Cycloaddition processes are classified with respect to three facets of the reaction: (1) the number of electrons of each unit participating in the cycloaddition, (2) the nature of the orbitals undergoing change (π or σ), and (3) the sterochemical mode of cycloaddition (*supra* or *antara*).

supra   *antara*

We denote the number of electrons in each unit by numbers and the nature of the orbitals by a prefix (π or σ). The stereochemical mode is given by a subscript (*s* or *a*), which indicates whether the addition occurs in a *supra* or *antara* mode on each unit.

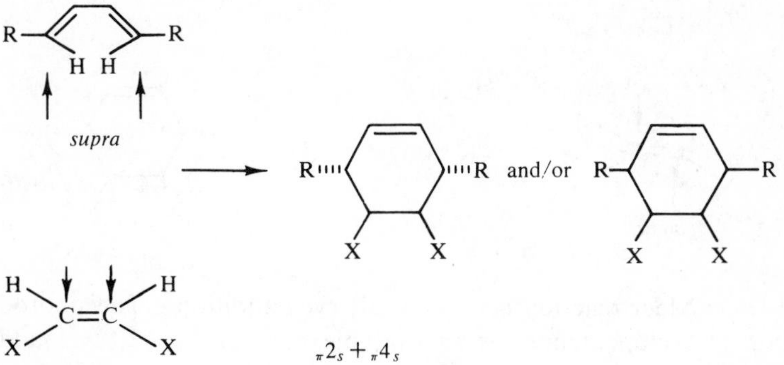

$R\text{''''}$)''''R and/or $R-$ $-R$

$_\pi2_s + {}_\pi4_s$

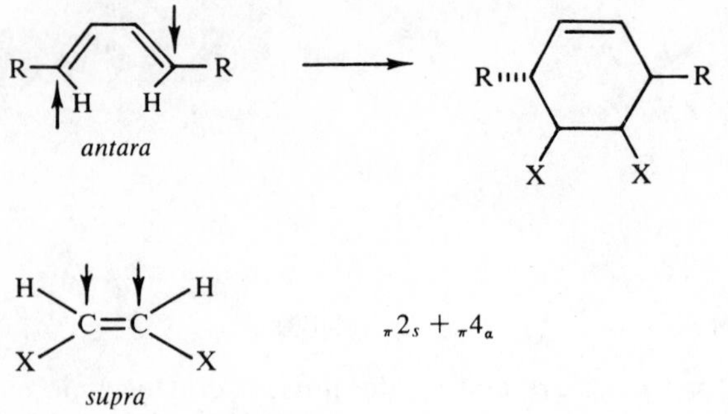

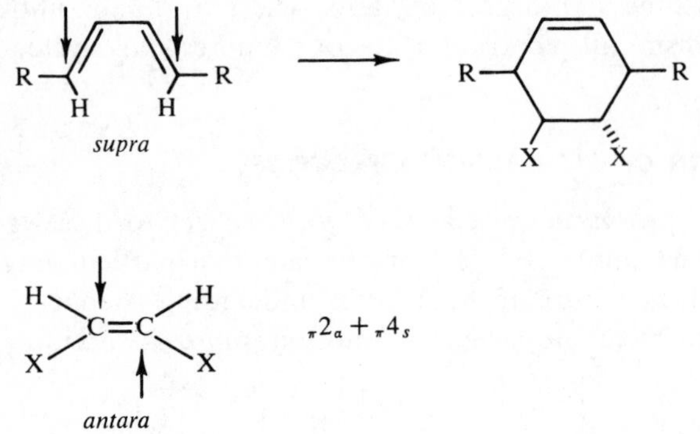

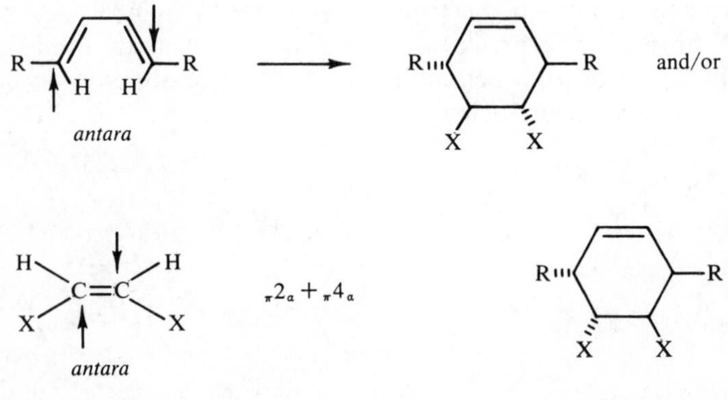

$_\pi 2_s + _\pi 4_a$

$_\pi 2_a + _\pi 4_s$

and/or

$_\pi 2_a + _\pi 4_a$

The Diels-Alder reaction is a $_\pi 2_s + _\pi 4_s$ cycloaddition. The photodimerization of cyclopentenone is an example of a $_\pi 2_s + _\pi 2_s$ cycloaddition.

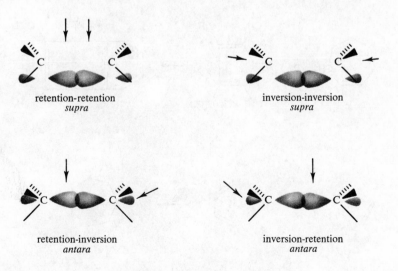

$_{\pi}2_s + _{\pi}4_s$

$_{\pi}2_s + _{\pi}2_s$

In cases where σ orbitals are involved, the meaning of the terms *supra* and *antara* are defined on the basis of retention or inversion at the two carbons involved. Retention (or inversion) at both carbon atoms is designated *supra*, whereas retention at one carbon and inversion at the other is designated *antara*, as illustrated in the following hypothetical reactions.

retention-retention
supra

inversion-inversion
supra

retention-inversion
antara

inversion-retention
antara

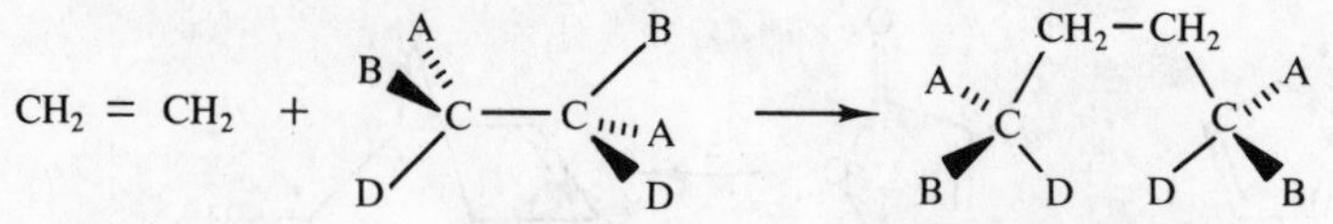

$_\pi 2_s + _\sigma 2_s$

retention-retention
supra

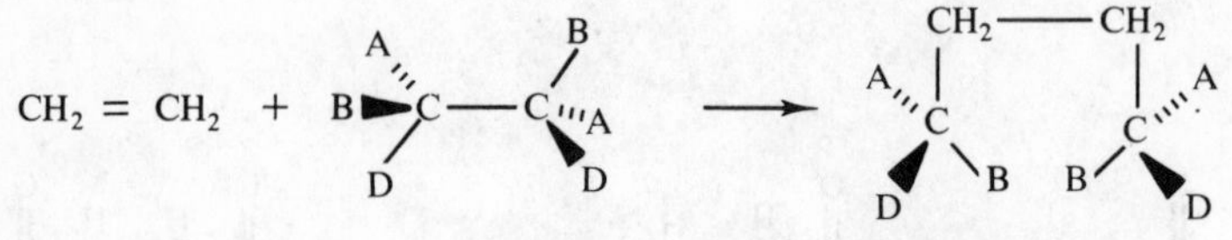

$_\pi 2_s + _\sigma 2_s$

inversion-inversion
supra

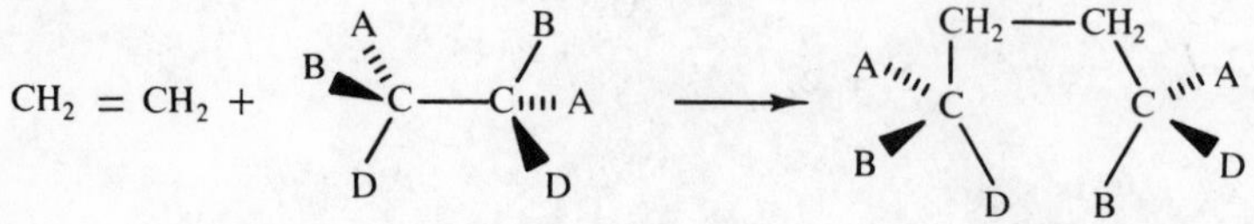

$_\pi 2_s + _\sigma 2_a$

retention-inversion
antara

Several examples of cycloaddition are given below with the proper classification.

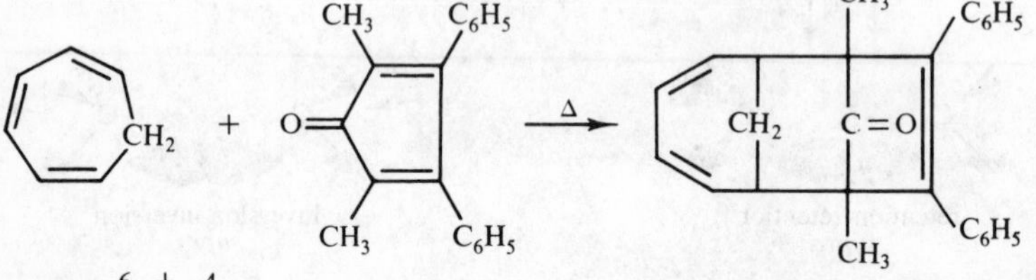

$_\pi 6_s + _\pi 4_s$

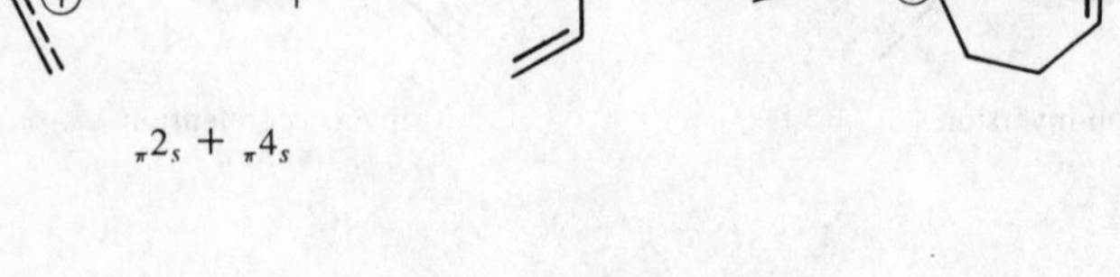

$_\pi 2_s + _\pi 4_s$

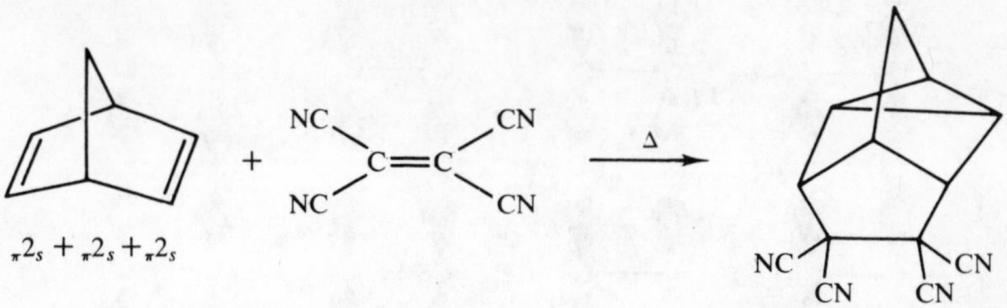

$_\pi2_s + {}_\pi2_s + {}_\pi2_s$

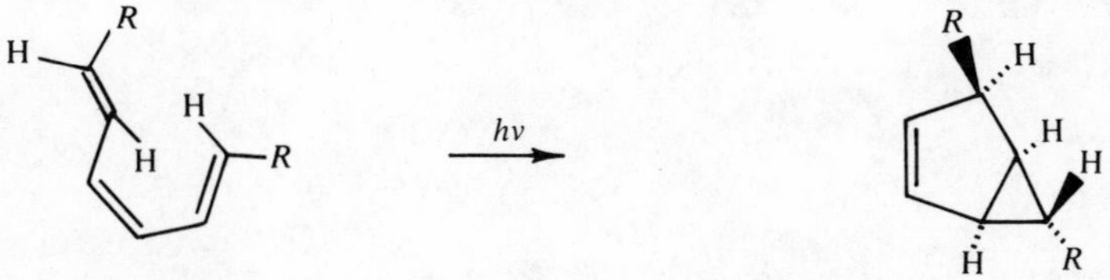

$_\pi2_a + {}_\pi4_s$

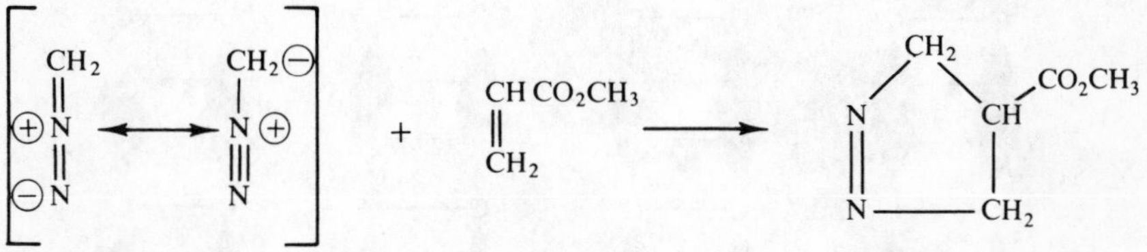

$_\pi2_s + {}_\pi4_s$

7.3 ORBITAL SYMMETRY AND CYCLOADDITION

Orbital symmetry arguments make useful predictions about concerted cycloaddition processes. Consider the $_\pi2_s + {}_\pi2_s$ cycloaddition of ethylene molecules in parallel planes approaching each other vertically. This system contains vertical and horizontal planes of symmetry σ_v and σ_h, which are

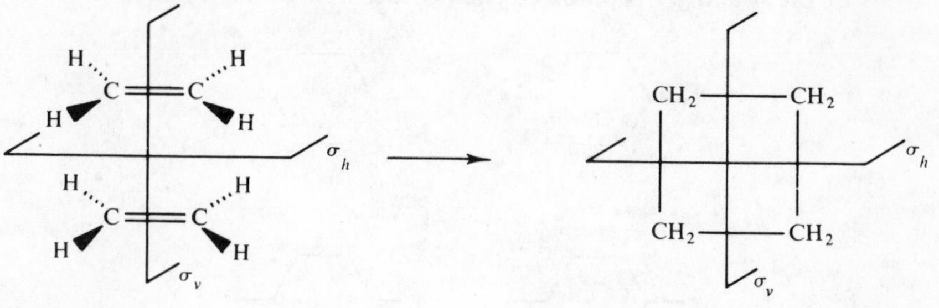

useful in characterizing the orbitals of the interacting ethylenes and the

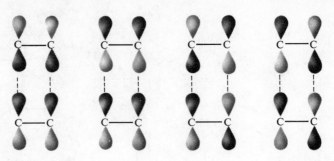

Symmetry of interacting ethylene π orbitals

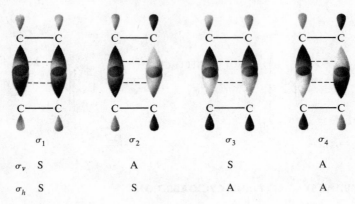

	σ_1	σ_2	σ_3	σ_4
σ_v	S	A	S	A
σ_h	S	S	A	A

Symmetry of interacting cyclobutane σ orbitals

product cyclobutane. The orbitals of the interacting ethylenes are the result of forming bonding and antibonding combinations of the π and π^* orbitals of the ethylenes. The interacting σ orbitals are similar combinations of the σ and σ^* orbitals of each of the new bonds formed in cyclo-

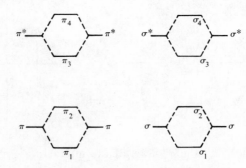

butane. We can now construct a correlation diagram for the $_\pi2_s + {}_\pi2_s$ cycloaddition.

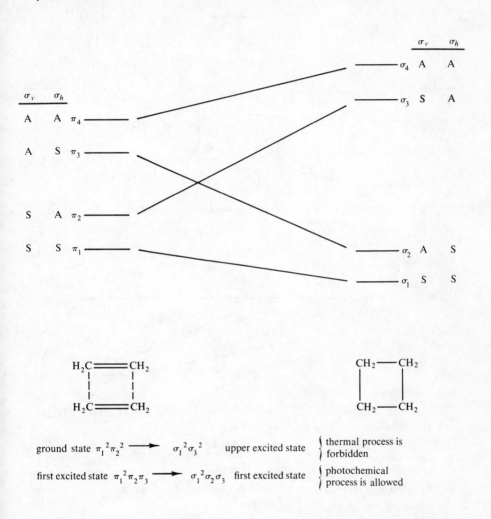

ground state $\pi_1^2\pi_2^2 \longrightarrow \sigma_1^2\sigma_3^2$ upper excited state $\begin{cases} \text{thermal process is} \\ \text{forbidden} \end{cases}$

first excited state $\pi_1^2\pi_2\pi_3 \longrightarrow \sigma_1^2\sigma_2\sigma_3$ first excited state $\begin{cases} \text{photochemical} \\ \text{process is allowed} \end{cases}$

The ground state of the interacting ethylene system correlates with an upper excited state of the cyclobutane and this process is forbidden. The first excited state of the interacting ethylenes, however, correlates with the first excited state of the cyclobutane. The photochemical $_\pi2_{s\,+\,\pi}2_s$ cyclo-addition is thus an allowed process.

A similar correlation diagram can be constructed for cycloadditions such as the Diels-Alder reaction. In this case we have only the vertical plane of symmetry (σ_v) available. The ground state of the starting system

Ground State $\psi_1^2 \pi^2 \psi_2^2 \longrightarrow \sigma_1^2 \sigma_2^2 \pi^2$ Ground State	Thermal reaction allowed
First Excited State $\psi_1^2 \pi^2 \psi_2 \psi_3 \longrightarrow \sigma_1^2 \sigma_2 \pi^2 \sigma_3$ Upper Excited State	Photochemical reaction forbidden

correlates with the ground state of the product, and the thermal process is allowed. The excited state of the starting system correlates with an upper excited state of the product, and the photochemical process is forbidden.

The formalism developed in Chapter 6 can be used to generate selection rules for cycloaddition reactions. In the transition state for a $_\pi 2_s +$ $_\pi 2_s$ cycloaddition we have zero nodes and four electrons, and the photo-

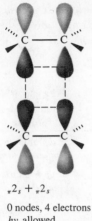

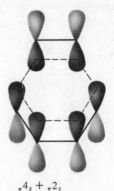

$_\pi 2_s + _\pi 2_s$

0 nodes, 4 electrons

hv allowed

$_\pi 4_s + _\pi 2_s$

0 nodes, 6 electrons

Δ allowed

chemical reaction is allowed. Cycloadditions of the $_\pi 4_s + _\pi 2_s$ type have transition states with zero nodes and six electrons and thus thermally allowed. If the addition is *antara* on one component, the array will have one node and the photochemical reaction will be allowed. Consider now the general case of a $_\pi m_s + _\pi n_s$ cycloaddition. There will be no nodes in

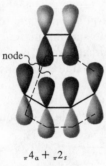

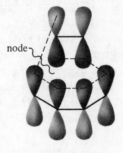

$_\pi 4_a + _\pi 2_s$

1 node, 6 electrons

hv allowed

$_\pi 4_s + _\pi 2_a$

1 node, 6 electrons

hv allowed

the array and $m + n$ electrons. If $m + n = 4q + 2$ (q = an integer), the

$$m\,\pi \overbrace{}^{\text{CH}_2} + \overbrace{}_{\text{CH}_2}^{\text{CH}_2} n\,\pi \longrightarrow (m-2)\pi \overbrace{}^{\text{CH}_2}_{\text{CH}_2} \overbrace{}^{\text{CH}_2}_{\text{CH}_2} (n-2)\pi$$

reaction will be thermally allowed. If $m + n = 4q$, the reaction will be photochemically allowed. For $_\pi m_a + _\pi n_a$, the array will still have zero nodes, and the same rules will hold. In cycloadditions that are *antara* on one component and *supra* on the other, there will be one node in the array. The selection rules for $_\pi m_a + {_\pi}n_s$ or $_\pi m_s + {_\pi}n_a$ are $m + n = 4q + 2$ pho-photochemical and $m + n = 4q$ thermal.

$m\pi + n\pi$	Allowed thermal process	Allowed photochemical process
$4q$†	*supra, antara* *antara, supra*	*supra, supra* *antara, antara*
$4q + 2$	*supra, supra* *antara, antara*	*supra, antara* *antara, supra*

†q is an integer.

It is interesting to note that $_\pi 2_s + {_\pi}2_a$ cycloadditions are allowed thermal processes and that the $_\pi 2_s + {_\pi}4_a$ and $_\pi 2_a + {_\pi}4_s$ cycloadditions are allowed photochemical processes. The predictions concerning *antara, antara* additions parallel those for *supra, supra* additions. Orbital overlap is much better for *supra, supra* addition so that *antara, antara* addition should be observed only in special systems.

A final warning concerning orbital symmetry arguments is appropriate at this point. It is clear why orbital symmetry arguments apply to ground state reactions (thermal processes). Ground state starting material is converted to ground state product, and orbital symmetry must be conserved. It is much less clear why the rules hold for excited state reactions (photochemical processes). In general, excited states (M^*) decay directly to ground state product (M') and do not go via product-excited states [$(M')^*$]. Nevertheless, orbital symmetry rules have been found experi-

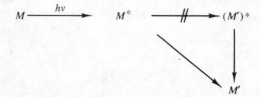

mentally to hold very well for both thermal and photochemical processes.

7.4 CONCERTED VS. NONCONCERTED CYCLOADDITION

Orbital symmetry arguments apply only when cycloaddition is a concerted process. We can imagine, for purposes of discussion, three modes of reaction: (1) concerted cycloaddition with equal bond formation in the transition state, (2) concerted cycloaddition with unequal bond formation in the transition state, and (3) nonconcerted cycloaddition with an intermediate. If we had only the first and third possibilities, we would have a clear distinction between concerted and nonconcerted processes.

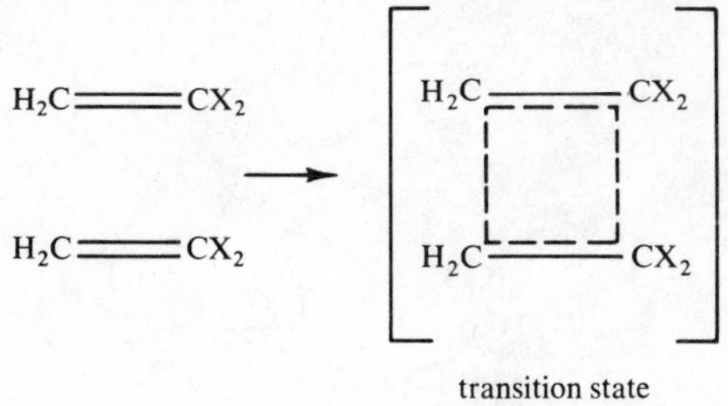

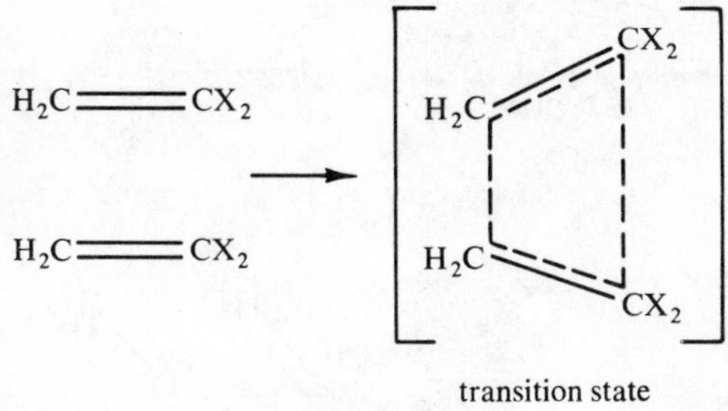

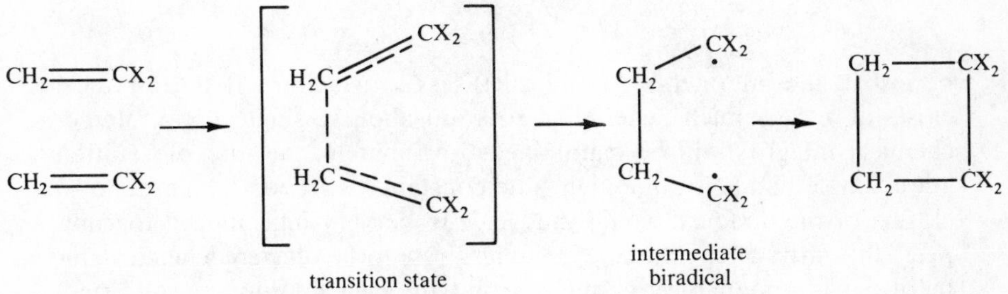

transition state

transition state

non-concerted process with a biradical intermediate

The second possibility, however, covers everything from a transition state with slightly unequal bonding to a transition state that looks very much like a biradical but has a weak interaction at the radical termini. There is a gradual change from concerted cycloaddition to nonconcerted cycloaddition. In making distinctions in this grey area chemists rely on an operational definition of a concerted process based on the stereochemical outcome of a given reaction. It is assumed that a concerted process will be stereospecific, whereas a nonconcerted process, which involves an intermediate, will not. The assumption concerning stereospecificity in a concerted process is sound. The assumption that stereochemical integrity will be lost in the intermediate biradical is less satisfactory. If the biradical closes at a rate much faster than rotation about a single bond, stereochemical integrity will be maintained. Fortunately, the rate of rotation about single bonds is quite high (rate constant $\sim 10^9$ sec^{-1}). The rate of closure of the biradical would thus have to be very high indeed to compete efficiently. If we observe complete retention of stereochemical integrity, it is reasonable to conclude that the cycloaddition is concerted. If we observe some loss of stereochemical integrity, it is reasonable to conclude that at least part of the reaction occurs via a nonconcerted cycloaddition.

7.5 $_\pi 2 + _\pi 2$ CYCLOADDITION

The orbital symmetry prediction that $_\pi 2_s + _\pi 2_s$ cycloaddition should be a photochemical process is strongly supported by experience. The examples available so far are primarily a combination of two ethylene units to form cyclobutanes. These reactions have found much use in synthesis of natural products and highly strained systems.

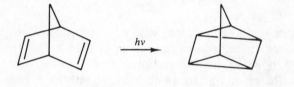

Photochemical cycloadditions raise several mechanistic questions. For example, what effect does the multiplicity of an excited state have on the reaction? Can a triplet excited state undergo a concerted cycloaddition, or will it inevitably lead to a triplet biradical? Very little work has been done on the stereochemical features of the 2 + 2 cycloaddition and less on the effect of excited state multiplicity on retention or loss of stereochemical integrity. The photocycloaddition of *trans*-stilbene to olefins has been studied in some detail (see Chapter 5). The addition involves attack of the S_1 state of *trans*-stilbene on the olefin. The T_1 state of *trans*-stilbene undergoes *trans-cis* isomerization, but it does not add to olefins. The stereochemical integrity of the *trans*-stilbene is maintained in the cyclo-

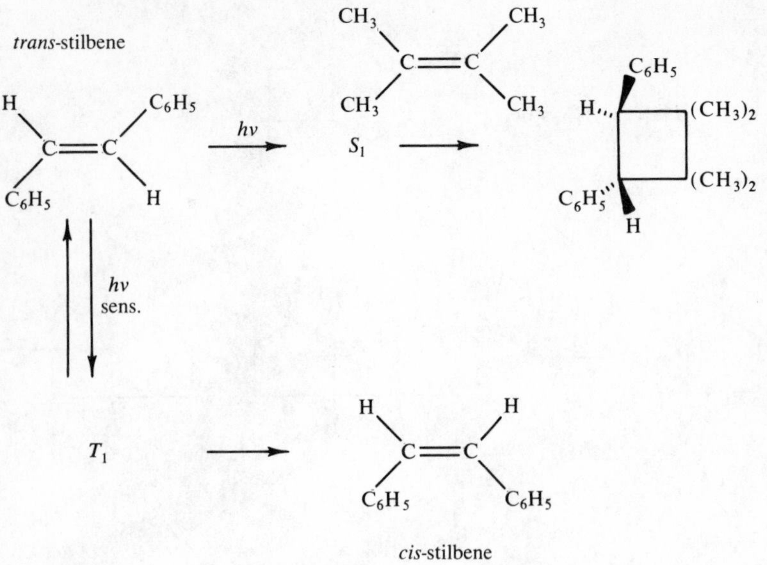

trans-stilbene

S_1

hv sens.

T_1

cis-stilbene

addition. Attempts to do a similar experiment with *cis*-stilbene have not been fruitful, because it isomerizes to *trans*-stilbene faster than it adds to olefins. It has been found, however, that *cis* and *trans* cinnamonitrile add to tetramethylethylene with retention of stereochemical integrity. The

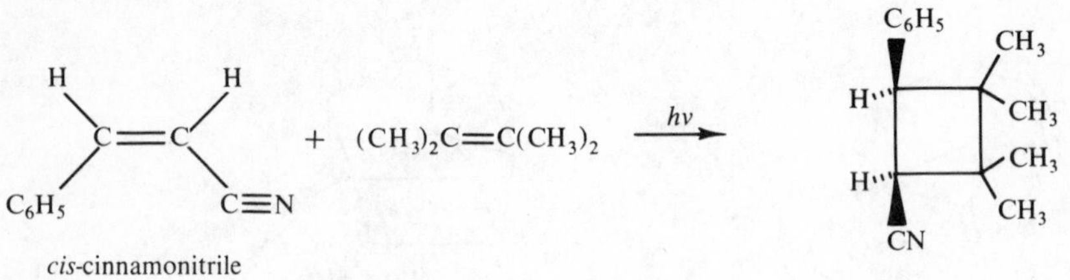

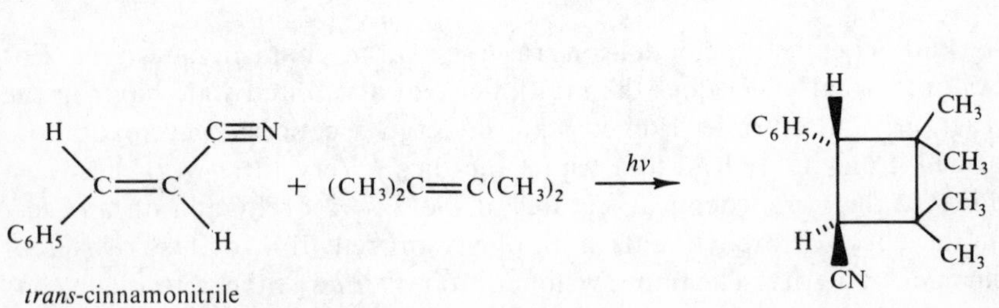

cis-cinnamonitrile

$+ \quad (CH_3)_2C=C(CH_3)_2 \quad \xrightarrow{hv}$

trans-cinnamonitrile

$+ \quad (CH_3)_2C=C(CH_3)_2 \quad \xrightarrow{hv}$

additions of *cis* and *trans* cinnamonitrile to olefins also involve singlet excited states. Stereochemical integrity is maintained in the excited component in these cycloadditions of singlet excited states.

We now turn to the stereochemical integrity of the ground state partner. Photoaddition of *trans*-stilbene to *cis* and *trans* 2-butene occurs with
retention of stereochemical integrity on both components. It is thus

reasonable, on the basis of our operational definition, to conclude that
the addition of the S_1 state of *trans*-stilbene to olefins is a concerted
process.†

The addition of *trans*-stilbene to tetramethylethylene proceeds by way
of an exciplex. The exciplex reverts to S_1 *trans*-stilbene and olefin and
demotes to adduct.

† If a biradical is involved, it must have a very short lifetime ($<10^{-9}$ sec).

$$S_t^* + h\nu \longrightarrow {}^1S_t$$

$${}^1S_t \xrightarrow{\ k_d\ } S_t$$

$${}^1S_t \xrightarrow{\ k_f\ } S_t + h\nu_f$$

$${}^1S_t \xrightarrow{\ k_i\ } S_{cis}$$

$${}^1S_t \xrightarrow{\ k_{ic}\ } {}^3S$$

$${}^1S_t + 0 \underset{k_{-e}}{\overset{k_e}{\rightleftharpoons}} [{}^1S_t \ldots 0] \quad \text{exciplex}$$

$$[{}^1S_t \ldots 0] \xrightarrow{\ k_a\ } \text{adduct}$$

It is possible from fluorescence quenching (see Chapter 5) and plots of $1/\Phi_a$ vs. $1/[O]$ to obtain values (10^9 1 mole^{-1}sec^{-1}) for the expression $k_e[k_a/(k_a + k_{-e})]$, i.e., the product of the rate constant for exciplex formation (k_e) and the fraction of exciplex that goes on to adduct $[k_a/(k_a + k_{-e})]$. The fraction of exciplex that goes on to adduct increases with decreasing temperature, and this leads to a large increase in quantum efficiency of cycloaddition as the temperature is lowered.

The sensitized (triplet) addition of cyclopentadiene to either *cis* or *trans* 1,2-dichloroethylene gives the same mixture of products and shows complete loss of stereochemical integrity. This is clearly a process that proceeds through a triplet biradical intermediate. It is not obvious, however, that all triplet cycloadditions proceed through triplet biradicals. The addition of 4,4-dimethyl-2-cyclohexenone to 1,1-dimethoxyethylene is known to be a triplet cycloaddition. In this addition the highly strained

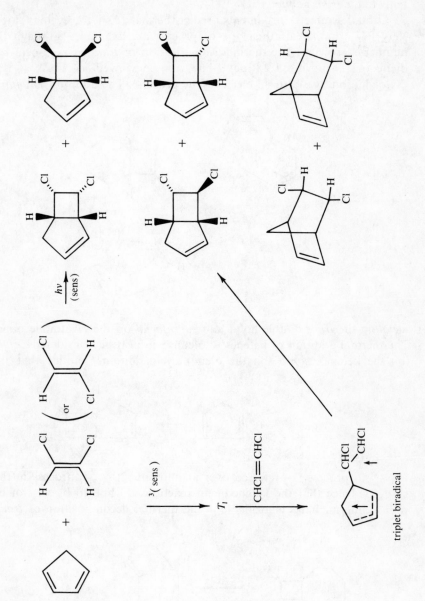

trans adduct is formed in higher yield than the more stable *cis* adduct. It would be quite surprising if a triplet biradical were to close preferentially to the *trans* adduct.

Orbital symmetry arguments predict thermal $_\pi 2_a + _\pi 2_s$ cycloaddition. Very few examples of this type of process are available, although the number is growing. The thermal isomerization of *trans,cis*-cycloocta-1,3-diene to *cis*-bicyclo[4.2.0]oct-7-ene can be viewed as a $_\pi 2_s + _\pi 2_a$ cycloaddition (it is also an electrocyclic process). The addition is *antara-*

facial on the *trans*-double bond and *suprafacial* on the *cis*-double bond. The thermal addition of ketenes to olefins can also be considered *antara* (on the ketene), *supra* (on the olefin) cycloadditions. Notice that the

$$H_2C = C = O$$
ketene

$_\pi 2_a + _\pi 2_s$ process is preferred over an allowed $_\pi 2_s + _\pi 4_s$ process in this case. Evidence that the ketene-olefin reactions do not go by way of biradical intermediates is available in the thermal decomposition of *trans-*

trans bicyclo [6.2.0]-
decan-9-one

bicyclo[6.2.0] decan-9-one to *trans*-cyclooctene and ketene. The thermal decomposition should follow the same correlation diagram as the addition.

Thermal addition of highly halogenated olefins to other olefins or dienes often gives halogenated cyclobutanes. These reactions have been shown to go through biradical intermediates. Addition of 1,1-dichloro-2,2-difluoroethylene to *trans,trans*-2,4-hexadiene gives two adducts. The presence of the adduct with *cis* methyl and propenyl groups shows that this reaction does not maintain stereochemical integrity and suggests that both products are formed via a biradical intermediate. Retention of the *trans* geometry in the propenyl group of both products is anticipated,

since allyl radicals are known to maintain their geometry under moderate conditions.

Formal $_\pi2 + _\pi2$ thermal cycloadditions that involve ionic intermediates are also known. The addition of *p*-methoxystyrene to tetracyanoethylene provides an example. The cyano groups stabilize the negative charge, and the *p*-methoxyphenyl group stabilizes the positive charge in the ionic intermediate. The reaction requires a powerful elec-

The chemical reaction scheme shows:

$$CH_3O-C_6H_4-CH=CH_2 \;+\; (NC)_2C=C(CN)_2 \;\longrightarrow\; CH_3O-C_6H_4-\overset{\oplus}{C}H-CH_2-\overset{\ominus}{C}(CN)_2-C(CN)_2-CN$$

ionic intermediate

$$\downarrow$$

(cyclobutane product with $CH_3O-C_6H_4$ substituent, ring bearing $(NC)_2$ and $(CN)_2$ groups)

tron releasing group in the phenyl ring and shows large rate enhancements on changing from nonpolar to polar solvents.

7.6 $_\pi 2 + _\pi 4$ CYCLOADDITIONS

Orbital symmetry arguments predict thermal $_\pi 2_s + _\pi 4_s$ cycloaddition and photochemical $_\pi 2_a + _\pi 4_s$ or $_\pi 2_s + _\pi 4_a$ cycloaddition. We shall discuss two thermal $_\pi 2 + _\pi 4$ cycloadditions, the Diels-Alder reaction and 1,3-dipolar addition, in some detail. Before turning to these important reactions, however, we shall consider one example of $_\pi 2_a + _\pi 4_s$ photo-

(reaction scheme: trans,cis,trans-1,3,5-hexatriene with R, H substituents $\xrightarrow{hv}$ bicyclo[3.1.0]hex-2-ene product with R, H stereochemistry shown)

chemical process. Irradiation of certain *trans,cis,trans*-1,3,5-hexatrienes leads to bicyclo[3.1.0]hex-2-enes with the stereochemistry shown. This transformation can be viewed as a $_\pi 2 + _\pi 4$ cycloaddition that goes *antara* on the olefin and *supra* on the diene. Atoms 1–4 comprise the diene

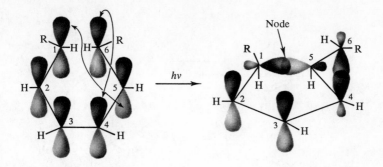

system and atoms 5 and 6 the olefin. The orbitals in the product have one node, as is expected for a photochemically allowed process in a six-electron system.

7.7 DIENE COMPONENT OF THE DIELS-ALDER REACTION

We shall now consider various aspects of reactivity and stereochemistry of the Diels-Alder reaction. Let us look first at the *diene* component of the reaction. The two double bonds must be conjugated so that a smooth reorganization of the bonding electrons leading to a stable product is possible. Equally important, the diene must be able to assume a *cisoid* conformation, as shown for butadiene. Ordinarily an acyclic conjugated diene will exist as a mobile equilibrium of *cisoid* and *transoid* forms. Only the *cisoid* form can react; reaction through the *transoid* form would lead to a highly strained *trans*-cyclohexene derivative.

If we wish to deduce how reactive a particular substituted butadiene will be in the Diels-Alder reaction, we must consider first of all what effect the substituents will have on the *cisoid,transoid* equilibrium. If we introduce a *trans*-l-methyl group into butadiene, the reactivity is little affected. *cis*-1-Methylbutadiene, on the other hand, undergoes a Diels-Alder reaction slowly, because the concentration of the *cisoid* form is greatly reduced because of steric repulsion between the methyl and vinyl groups. *cis*-1-*t*-Butyl-1,3-butadiene and *cis,cis*-1,4-dimethylbutadiene, in which this type of repulsion is even greater, do not undergo the Diels-

Alder reaction at all.   A single bulky substituent at the 2-position of butadiene favors the *cisoid* form, and leads to an increase in reactivity. Large substituents at both the 2- and 3-positions again make the transoid form relatively the more stable form, and so decrease reactivity.

If the 2,3-positions of butadiene are connected by a small ring, then the two olefinic bonds are constrained to the *cisoid* form, and cycloaddition becomes rapid (a).   Similarly, only the *cisoid* form is possible if the two double bonds are both within the same small ring (homoannular

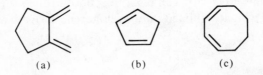

(a)   (b)   (c)

dienes). Cyclopentadiene (b) is among the most reactive dienes known. On the other hand, 1,3-cyclooctadiene (c) undergoes the Diels-Alder reaction only with difficulty, and molecular models show that its two double bonds are nearly at right angles to one another. If the double bonds of the diene are in different rings (heteroannular diene), then again only the *transoid* form is possible, and the Diels-Alder reaction is prohibited.

The electronic effect of a substituent on the diene is usually less profound than the conformational effect. Electronic effects in the diene are strongly coupled to electronic effects in the dienophile, and it is necessary to consider these effects together.

7.8 DIENOPHILE REACTIVITY

The two-atom unsaturated center that adds to the diene is known as the *dienophile*. It may be a carbon-carbon double or triple bond, and one or more heteroatoms may be present (R—C≡N, R—B=O, R—N=N—R, etc.). Some examples are given below. .

Electronic effects are more important than steric effects in determining the relative reactivity of dienophiles. Generally, the attachment of electron-attracting substituents (—C=O, —CN, —NO$_2$) to the dienophile center will increase the rate of reaction. Ethylene, for example, reacts slowly with cyclopentadiene. If one hydrogen is replaced by a cyano group, giving acrylonitrile, then reaction occurs readily. At 20° *cis*- and *trans*-1,2-di-cyanoethylene are about 100 times more reactive than acrylonitrile, and 1,1-dicyanoethylene is 45,000 times more reactive. Tricyanoethylene is 500,000 and tetracyanoethylene 43,000,000 times more reactive than acrylonitrile with cyclopentadiene as the diene. Similar large effects are noted upon replacement of ethylene hydrogen by an ester, ketone, and other electron-withdrawing groups. Even a double bond will act as a mild activating group, so that butadiene dimerizes thermally by a Diels-Alder reaction.

If we compare the reactivity of various substituted dienes with a highly electron-deficient dienophilic center like that in tetracyanoethylene, we find, as we might expect, that *electron-donating* substituents increase the rate of reaction, and electron-attracting substituents decrease it. *trans*-l-

Methoxybutadiene, for instance, is 1000 times more reactive toward tetra-cyanoethylene than is butadiene, whereas 2-chlorobutadiene is 500 times less reactive. Cyclopentadiene is 82,000 times more reactive than buta-diene toward tetracyanoethylene, yet hexachlorocyclopentadiene does not react at all with this dienophile. Most Diels-Alder reactions involve an electron-deficient dienophile and an electron-rich diene. It is possible, however, to have Diels-Alder reactions with exactly the opposite elec-tronic distribution, *viz.*, an electron-deficient diene and an electron-rich dienophile. Thus, hexachlorocyclopentadiene fails to react with tetra-cyanoethylene and reacts only slowly with maleic anhydride. It reacts more readily with unsubstituted olefins (cyclopentene), more readily yet with an enolether, and still more readily with styrene. *p*-Methoxystyrene (electron-donating substituent) is more reactive than styrene, which in turn is more reactive than *p*-nitrostyrene, just opposite to the order of reactivity found with more typical dienes. In summary, we may say that Diels-Alder reactions are susceptible to acceleration by electronic effects of substituent groups. The important thing is that the substituent elec-tronic effect be different in the two compounds, either electron-donating in the diene and electron-withdrawing in the dienophile (the more usual situation), or vice versa.

7.9 ORIENTATION EFFECTS IN DIELS-ALDER REACTIONS

If 2-methylbutadiene is allowed to react with methyl acrylate, two different Diels-Alder adducts could be formed, depending upon the rela-tive orientation of the diene and dienophile at the time of reaction.

Ordinarily a mixture of the two possible products will be formed.

7.10 CATALYSIS OF DIELS-ALDER REACTIONS

For many years it was thought that Diels-Alder reactions were not susceptible to catalysis. Recently it has been shown that powerful Lewis acids ($AlCl_3$, BF_3, $SnCl_4$) can greatly accelerate certain Diels-Alder reactions. Methyl vinyl ketone, for example, which adds to 2-methylbuta-diene in toluene only at 120°, does so at room temperature in the same solvent in the presence of stannic chloride. Undoubtedly, the catalyst complexes with the oxygen of the carbonyl group, increasing its electron-withdrawing ability and thus activating the double bond. Catalyzed Diels-

$$
\begin{array}{c}
\overset{\displaystyle O}{\underset{\displaystyle \parallel}{\text{CH}-\text{C}-\text{CH}_3}} \\
\underset{\displaystyle \text{CH}_2}{\parallel}
\end{array}
\quad + \quad \text{SnCl}_4 \quad \longrightarrow \quad
\begin{array}{c}
\overset{\ominus}{\overset{\displaystyle \text{OSnCl}_4}{\underset{\displaystyle \oplus}{\text{CH}-\text{C}-\text{CH}_3}}} \\
\underset{\displaystyle \text{CH}_2}{\parallel}
\end{array}
$$

Alder reactions have been carried out at temperatures as low as $-70°$. In addition to an increase in rate of reaction, the catalyst also seems to increase the orientational purity of the product, often leading to the selective formation of one isomer.

7.11 RETRO-DIELS-ALDER REACTION

Diels-Alder reactions are, in theory and usually in practice, reversible. Cyclopentadiene and its dimer, dicyclopentadiene, furnish a useful and instructive example. This cyclic diene is highly reactive as a diene and moderately so as a dienophile. On standing at room temperature it dimerizes in a day or two by a Diels-Alder reaction, forming dicyclopentadiene. This dimer can be purchased cheaply (it is formed in the

cracking of petroleum), but the monomeric diene must be freshly prepared before use. When the dimer is heated at 170–200°, a retro-Diels-Alder reaction occurs, and two molecules of cyclopentadiene are formed. The monomer is removed by distillation as it is formed and collected in an ice bath. In this way nearly all of the dimer may be reconverted to monomer.

The ease with which a given Diels-Alder adduct will revert to its components varies greatly, depending upon the relative stability of reactants and products. In some cases a mobile equilibrium exists at room temperature. This is the case for the Diels-Alder dimer of 2,5-dimethyl-3,4-diphenylcyclopentadienone. The large number of substituents causes steric repulsions in the dimer, while helping to stabilize the monomeric form. If the methyl groups are replaced by hydrogen, the dimer forms completely. On the other hand, tetraphenylcyclopentadienone is monomeric at room temperature.

The retro-Diels-Alder reaction has many practical synthetic applications. An unsaturated group may be protected while some reaction or reactions potentially destructive to it are being carried out elsewhere in the molecule. The double bond is then regenerated upon heating. Retro-Diels-Alder reactions may also be used in order to locate the position of substituents in cyclohexadiene rings. In this case a Diels-Alder reaction is carried out with dimethyl acetylene dicarboxylate. The resulting adduct

$$\text{(a)} \quad + \quad CH_2 = CH_2$$

(a)

(a) could also be formed, in theory, by the Diels-Alder reaction between ethylene and the benzene derivative shown. Although the forward reaction does not occur, its retro counterpart does so readily, since the product gains the resonance energy of benzene.

Since retro-Diels-Alder reactions may be carried out in the vapor phase in the absence of all solvents and catalysts, they may be used to prepare some reactive compounds that are difficult to prepare by other methods.

7.12 STEREOCHEMISTRY OF THE DIELS-ALDER REACTION

In a Diels-Alder reaction between a diene and alkene, four trigonal centers in reactants become tetrahedral in the product. As a consequence, a large number of stereoisomeric products might be formed. The great power of the Diels-Alder reaction lies in the fact that it is ordinarily possible to predict, from the structure of the reactants, which isomer will be formed, and, conversely, to synthesize a desired stereo-isomer by proper choice of reactant. The Diels-Alder reaction is highly stereospecific, and stereospecific in a predictable fashion.

Within each diene or dienophilic component, substituents will retain the stereochemical relationship to one another in the product that they had in the reactants. As an example, diethyl maleate, in which the carbethoxy groups are *cis*, reacts with butadiene to give diethyl *cis*-tetra-hydrophthalate, whereas fumarate esters (*trans*-carboxyl groups) give *trans*-products. The stereochemical control here is absolute; i.e., it is not a

matter of starting with the *cis* and getting a mixture in which the *cis* pre-dominates. It is, in fact, not possible to detect the other isomer by tech-niques that would pick up one part in a million if it were there. This is the best evidence that bonding occurs at both ends of the alkene at the same time, i.e., that the reaction is *concerted*.

The product stereochemistry is just as precisely controlled by the geometry of the diene. Groups 1,4 on the outside of the *cisoid* diene (*trans,trans*-1,4-substituents) end up *cis* to one another on one side of the cyclohexene ring; groups on the inside (cis,cis-1,4-substituents) end up *cis* to one another on the opposite side of the ring.

Finally, we need to consider the stereochemistry in the product of substituents on the diene relative to those on the dienophile. If *trans*-1,3-pentadiene reacts with acrylonitrile, we know that in the product the methyl and cyano groups could be either *cis* or *trans*. We find, in fact, that the *cis* product is formed predominantly from *trans*-1,3-pentadiene

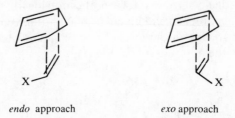

and the *trans* product from *cis*-1,3-pentadiene.

A study of a large number of such reactions has led to a generalization that has become known as Alder's rule of *endo* addition. The rule can be illustrated most easily by considering the reaction of cyclopentadiene with a substituted ethylene. As the planar ethylene approaches the planar diene, it can do so with the substituent X oriented *endo, over the* four carbons of the diene system, or *exo*, with X away from the diene. From

endo approach   exo approach

observation of experimental results, it has become obvious that *endo* addition is preferred. The reason for preferred *endo* approach is thought to be a favorable interaction between the orbitals of the substituent and those of the internal atoms of the diene. Application of the *endo* rule to the reactions of *trans* and *cis* 1,3-pentadiene with acrylonitrile shows the origin of the stereospecific product formation.

cis

trans

In general, there is only a modest preference for *endo* addition. The reaction of cyclopentadiene with methyl acrylate gives predominantly *endo* product, but both isomers are formed. The *exo* product is thermo-

endo   *exo*

dynamically more stable than the *endo* product. If either adduct is heated, an equilibrium will be established between cyclopentadiene, methyl acrylate, *endo* product, and *exo* product. At equilibrium there will be more *exo* product than *endo* product. Reactions run under conditions that achieve equilibrium are thermodynamically controlled. Product composition in reactions that are run under conditions that do not·equilibrate the products will depend on the rate at which the products are formed. Such reactions are kinetically controlled. In the Diels-Alder reaction *endo* products are formed faster than *exo* products, and kinetically controlled reactions result in predominantly *endo* products. Thermodynamically controlled reactions favor the more stable *exo* products. Lower reaction temperature (made possible in many cases by the use of Lewis acid catalysts) favors kinetic control.

7.13 1,3-DIPOLARADDITIONS

In the Diels-Alder reaction a four-atom and a two-atom component react to form a six-membered ring. In 1,3-dipolar additions a three-atom and a two-atom component form a five-membered ring. Both reactions are $_\pi 2_s + {}_\pi 4_s$ cycloadditions. In order for the product to be uncharged, at least one of the components must bear formal charges. Ordinarily a *1,3-dipole* reacts with an unsaturated center (*the dipolarophile*). Simple examples are the addition of phenylazide or diazomethane to a double bond. The 1,3-dipolar species may be relatively stable molecules like these,

or they may be highly reactive intermediates that must be generated *in situ*. Diphenylnitrilimine is an example of the latter type. It is prepared in the presence of the dipolarophile by heating a tetrazole to 160–180°. The reaction is a convenient one for the synthesis of Δ^2-pyrazolines. The

structures of a number of 1,3-dipolar molecules that have been used in 1,3-dipolar addition reactions are given in the following table.

1,3-dipole	name
$N{\equiv}\overset{\oplus}{N}-\overset{\ominus}{C}R_2 \longleftrightarrow \overset{\oplus}{N}{=}N-\overset{\ominus}{C}R_2$	diazo compounds
$N{\equiv}\overset{\oplus}{N}-\overset{\ominus}{O} \longleftrightarrow \overset{\oplus}{N}{=}N-\overset{\ominus}{O}$	nitrous oxide
$N{\equiv}\overset{\oplus}{N}-\overset{\ominus}{N}-R \longleftrightarrow \overset{\oplus}{N}{=}N-\overset{\ominus}{N}-R$	azides
$\overset{\ominus}{O}-\overset{\oplus}{N}{\equiv}C-R \longleftrightarrow \overset{\ominus}{O}-\overset{\oplus}{N}{=}C-R$	nitrile oxides
$\overset{\ominus}{O}-\overset{\oplus}{O}{=}O \longleftrightarrow \overset{\ominus}{O}-\overset{\oplus}{O}-O$	ozone
$R-\overset{\ominus}{N}-\overset{\oplus}{N}{\equiv}C-R' \longleftrightarrow R-\overset{\ominus}{N}-N{=}\overset{\oplus}{C}-R'$	azomethine imines
$\overset{\ominus}{O}-\overset{\oplus}{N}{=}CR_2' \longleftrightarrow \overset{\ominus}{O}-\overset{\oplus}{N}-CR_2'$ $\qquad$ R $\qquad\qquad$ R	nitrones

The 1,3-dipolar compounds shown above all have reasonable stability, even if they cannot always be isolated, because all three atoms can at least share an octet of electrons. For example, the following resonance structures may be written for a nitrone. It is possible, however, to conceive of 1,3-dipolar molecules for which octet structures cannot be

$$\overset{\ominus}{O}-\overset{\cdot\cdot}{\underset{\underset{R}{|}}{N}}-\overset{\oplus}{\underset{\underset{R'}{|}}{C}}-R' \quad\longleftrightarrow\quad \overset{\ominus}{O}-\overset{\oplus}{N}{=}\underset{\underset{R'}{|}}{C}-R' \quad\longleftrightarrow\quad \overset{\oplus}{O}{=}\overset{\ominus}{N}-\underset{\underset{R'}{|}}{C}-R'$$
$$\qquad\quad\;\, R \quad R'$$

written. These are expected to be of high energy and will be correspondingly more difficult to detect by cycloadditions. One of the few such systems to be studied is the ketocarbene shown below. Only in special cases

can the cycloaddition reaction compete with other possible modes of reaction of these intermediates.

1,3-Dipolar cycloadditions are stereospecific. Diazomethane adds to dimethyl dimethylfumarate and dimethyl dimethylmaleate to give isomeric pyrazolines. It is interesting to compare the relative reactivities

of a series of olefins as dipolarophiles in 1,3-cycloadditions and as dienophiles in the Diels-Alder reaction. In general, the overall pattern of
reactivity is similar, with electron-withdrawing groups attached to the
double bond increasing the rates of reaction. Thus maleic anhydride reacts
with diphenyldiazomethane 4000 times more rapidly than does styrene,
and ethyl acrylate reacts 500 times more rapidly. 1,3-Dipolar additions
show very little effect of solvent on the rates of addition, indicating that
the reaction is highly concerted with little charge separation in the transition state.

In many 1,3-additions, questions of orientation arise. The product
formed appears to be determined mainly by steric effects. Thus, reaction
of phenylnitrile oxide with styrene leads to a Δ^2 isoxazoline with the two
phenyl groups as far away from one another as possible.

7.14 PROBLEMS

1. Predict the stereochemistry of the following concerted reactions.

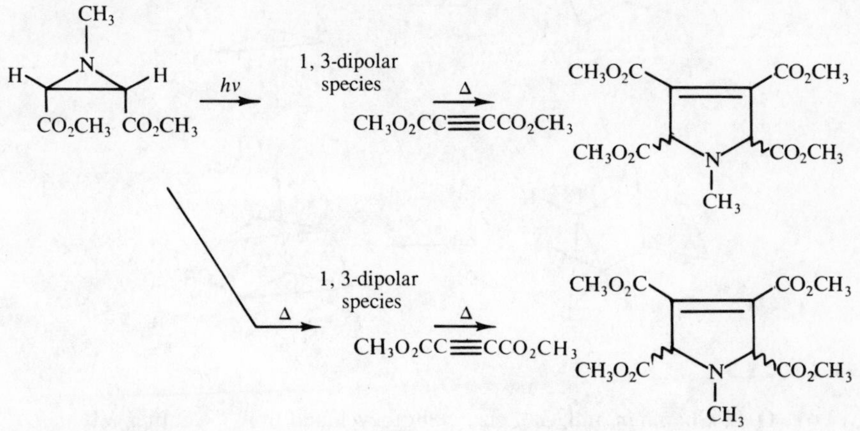

2. Write a mechanism for the following transformation which involves two thermally allowed concerted reactions.

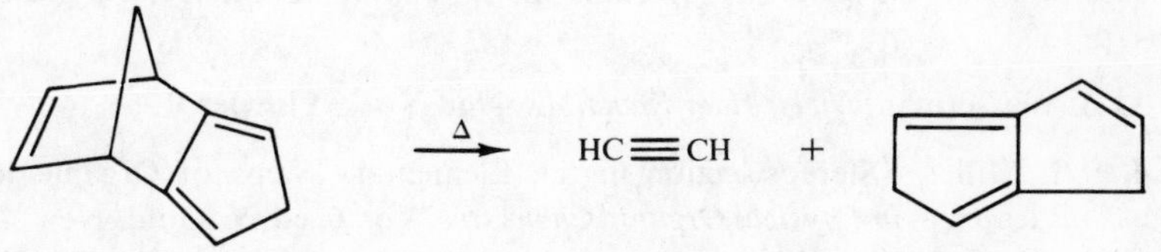

3. Write a mechanism for the following reaction.

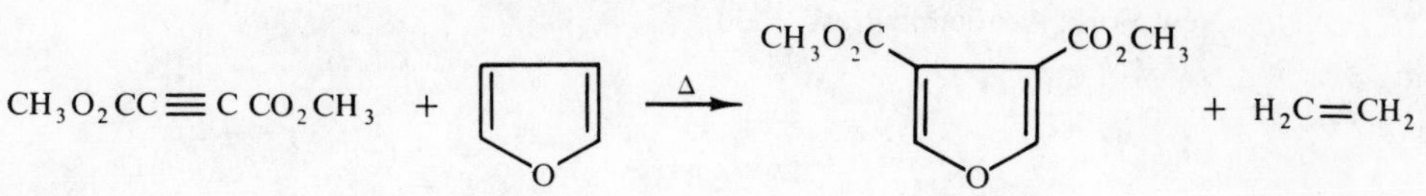

4. Write a mechanism for the following reaction.

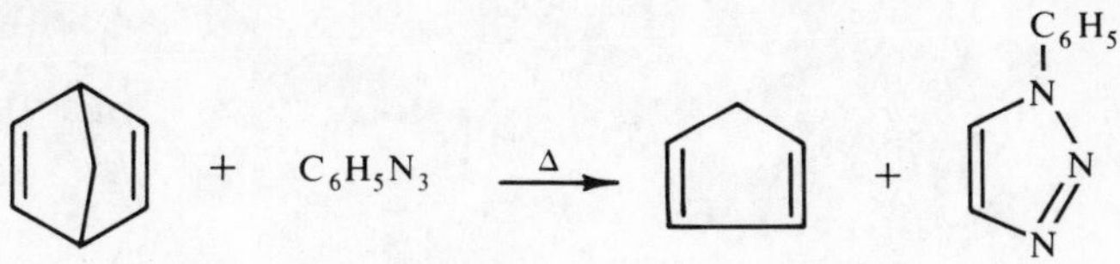

5. Consider the two $_\pi 2_s + _\sigma 2_s$ cycloadditions shown below. One involves retention-retention and the other inversion-inversion on the carbon atoms of the sigma bond. Indicate which process is involved for each compound. Which reaction do you expect to be more facile? Why?

REFERENCES

1. O. L. Chapman and G. Lenz, "Photocycloaddition," Chapter VII in *Organic Photochemistry*, Vol. 1, ed. O. L. Chapman. New York: Dekker, 1967.

2. R. Huisgen, R. Grashey, and J. Sauer, "Cycloaddition Reactions of Alkenes," in *The Chemistry of Alkenes*, ed. S. Patai. New York: Interscience, 1964, p. 739.

3. A. Wasserman, *Diels-Alder Reactions*. New York: Elsevier, 1965.

4. S. I. Miller, "Stereoselection in the Elementary Steps of Organic Reactions," in *Advances in Physical Organic Chemistry*, Vol. 6, ed. V. Gold. New York: Academic Press, 1968, p. 185.

5. R. B. Woodward and R. Hoffmann, *The Conservation of Orbital Symmetry*. New York: Academic Press, 1970.

Index